Lecture Notes in Mathematics

ISBN 978-3-540-04932-6 © Springer-Verlag Berlin Heidelberg 2006

Karl W. Gruenberg

Cohomological Topics in Group Theory

Comments and Corrections

p. 88, end of §6.1. Add the sentence "This construction is independent of the chosen I for A ".

p. 119. Chapter 8 has a natural continuation in Chapter 2 of Robert Bieri's Queen Mary College Mathematics Notes *Homological Dimension of Discrete Groups*, 2nd. edition (1981).

p. 121, line −6. After "all A " insert "and all $q > k$ ".

p. 129. A theorem by C.T.C. Wall and J.-P. Serre gives a cohomological condition that implies conjugacy conclusions on the finite subgroups. This gives an attractive proof of an old result about one-relator groups: *if G is one-relator with maximal finite cyclic subgroup C, then any finite subgroup of G is conjugate to a subgroup of C.* Neither Serre nor Wall published the result; but a recent good account is in the Appendix (p. 597) by C. Scheiderer to the paper by P. Lochak and L. Schneps in Inventiones 127 (1997).

p. 144. In exercise 5, line 2, before "prove that" add the phrase "and $n > 0$ or at least one $s_i = \infty$, or $m > 3$ ".
There is much more about polyhedral groups (Fuchsian groups of genus 0) in L.L. Scott *Matrices and cohomology*, Annals Math. 105 (1977) 473–492.

p. 155. The strict inequality ($\star$) is always true: cf. Bieri's Queen Mary College Mathematics Notes §8.3 (p. 120).

p. 173, Exercise. Magnus' result has very recently been generalised to its best possible form by J.S. Wilson (*On growth of groups with few relators*, Bull. London Math. Soc. 36 (2004) 1–2), based on earlier work of N.S. Romanovskii. The result is the following: *If G has a presentation with $r + s$ generators and r relations with $s > 0$, and S is an arbitrary generating set of G, then some subset of s elements of S freely generates a free subgroup of G.*

p. 175, end of §8.11. For more examples of groups with trivial cohomological dimension 0, cf. P. de la Harpe and D. McDuff in *Acyclic groups of automorphisms*, Comment. Math. Helvetici 58 (1983) 48–71. (They call a group G acyclic if $tcd\, G = 0$.)

p. 202. In line 3 of the definition add "and $A_1 \neq 0$ ".
Comment: $(0,0)$ is a projective pair in $\mathcal{Q}_G$ if, and only if, G is a free group.

p. 212, line −5. " ... and $(A, co(A|E))$ is ... "

p. 216, Problem. $\mathbb{Q}$ has no minimal projectives. Cf. J.S. Williams *Nielsen equivalence of presentations of some solvable groups*, Math. Z. 137 (1974) 351–362.

p. 227. In the definition of Heller module, it is better to allow KG-projectives to be Heller modules.
Proposition 4 is incorrect as it stands. KG must be assumed to have the following further property: if U, V are KG-lattices, then $(U \oplus V)' \simeq U' \oplus V'$. (If we allow KG-projectives to be Heller modules, then the conclusion of Proposition 4 is just that C is a Heller module if, and only if, C' is indecomposable.)
Comment. That the first property of KG (in Proposition 4) is insufficient for the conclusion is due to the existence of groups G with minimal relation modules $\bar{R}$ having the property that $\bar{R}^{(n)}$ has $\mathbb{Z}G$ as a direct summand for some $n > 1$. If n is the smallest such number and $K = \mathbb{Z}_{(G)}$, then $U = \bar{R}^{(n-1)}_{(G)}$ and $V = \bar{R}_{(G)}$ have $U' = U$ and $V' = V$, but $(U \oplus V)' < U \oplus V$. Cf. J.S. Williams *Trace ideals of relation modules of finite groups*, Math. Z. 163 (1978) 261–274.

p. 228, Proposition 5. The proof really gives a bit more: *C is a Heller module if, and only if, A is a Heller module.*

p. 229. Misprint in line 7: $KG \simeq K^*$

p. 230. Replace "Problem" with: "Note that $\mathbb{Z}$ is a Heller module if, and only if, $\mathbb{Z}_{(G)}$ is a Heller module."
Comment. The question which groups have $\mathbb{Z}$ as Heller module has led to a substantial body of work. It is now known (modulo CFSG) that $\mathbb{Z}$ is a Heller module for G if, and only if, the prime graph of G is connected: cf. J.S. Williams *Prime graph components of finite groups*, J. Alg. 69 (1981) 487–513. My survey in London Math. Soc. Lecture Note Series 36 (1979) discusses the links between prime graphs and Heller properties. These and related questions were taken impressively further (for not necessarily finite groups) by P.A. Linnell in *Decomposition of augmentation ideals and relation modules*, Proc. London Math. Soc. 47 (1983) 83–127. In another direction, the representation-theoretic significance of Heller-type properties for general lattices is studied by S.N. Aloneftis in *Decomposition modulo projectives of lattices over finite groups*, J. Alg. 223 (2000) 1–14.

p. 237. After line 1 add: "Strictly, $K_{(G)}$ is undefined if $|G|$ is a unit in K. When this is so, set $K_{(G)}$ to be the field of fractions of K."

p. 240. Definition should read: "A ring L is called semi-local if $L/Jac(L)$ has descending chain condition on right ideals (and so is semi-simple)".

Exercise 1 should read: "The commutative ring L is semi-local if, and only if, L has only a finite number of maximal ideals."

p. 242. The last sentence of the Corollary should read: "Then KG has the first of the two properties required in Proposition 4 of §10.2 (p. 227). (Cf. the comment above for p. 227.)"

p. 243. Delete the line immediately before equation (3) (i.e., "and only if").

p. 252. For Theorem 1 one needs to know that minimal projectives exist. They do: this is a consequence of the fact that if KG is semi-perfect, then every finitely generated KG-module satisfies the ascending chain condition on projective direct summands (use (i) KG/J is semi-simple, where $J = Jac(KG)$, and (ii) if P is KG-projective, then $P/PJ \neq 0$; cf. Bass' paper cited on p. 274).

p. 258, line 4. A should be A^G.

p. 264, line 4. Theorem 5 does not remain true over $\mathbb{Z}$. Here is an almost trivial example: Let $G = \mathbb{Z}/p\mathbb{Z}$, where p is a prime > 3; let $\varphi, \psi : \mathbb{Z} \twoheadrightarrow \mathbb{Z}/p\mathbb{Z}$ be the homomorphisms $\varphi(1) = 1$, $\psi(1) = 2$. Both give minimal projective objects but they are not isomorphic.
On a distinctly non-trivial level, modules in minimal projectives need not be isomorphic. This was first proved by Dyer and Sieradski (J. Pure and Appl. Algebra 15 (1979) 199–217) and refined by P.J. Webb in *The minimal relation modules of a finite abelian group*, J. Pure and Appl. Algebra 21 (1981) 205–232. This paper is also relevant for Theorem 6 on this page.

p. 268, last line. This soon ceased to be an open problem. A free extension is minimal if, and only if, the relation module is core-equal (has no non-zero projective direct summand). This is known for large classes of groups (cf. my article in London Math. Soc. Lecture Note Series 36, 1979).

p. 270. In the statement of Theorem 9 delete "KG has the property of Proposition 4 of §10.2 (p. 227) and"
In the paragraph following Theorem 9 keep sentence 2 and delete the rest.
In the proof of Theorem 9 delete the first paragraph and retain the rest.

p. 271. The Corollary is false (since $\mathbb{Z}_{(G)}$ need not be Heller). This has led to a series of papers (and more errors! - but all finally resolved satisfactorily):

1. *Decomposition of the augmentation ideal and of the relation modules of a finite group*, Proc. London Math. Soc. 31 (1975) 149–166;
2. *Decomposition of the relation modules of a finite group*, J. London Math. Soc. 12 (1976) 262–266;
3. *The decomposition of relation modules: a corrrection*, Proc. London Math. Soc. 45 (1982) 89–96.

These 3 papers are by K.W. Gruenberg and K.W. Roggenkamp.

4. R.M. Guralnick and W. Kimmerle *On the cohomology of alternating and symmetric groups and decomposition of relation modules*, J. Pure and Appl. Algebra 69 (1990) 135–140.

p. 272. In the statement of Theorem 10, delete the reference to Proposition 4, §10.2.

p. 273, line 3. Minimal projectives need not be isomorphic: cf. the comments (above) to p. 264.
Concerning the various questions raised on this page, I mention two surveys of the theory published a few years after publication of this book: K.W. Roggenkamp *Integral representations and presentations of finite groups* Lecture Notes in Mathematics 744 (Springer) 1979; and my article in London Math. Soc. Lecture Note Series 36 (1979).
In the theorem stated at the end of the page, it is implicitly assumed that G is a p-group. This is a very special case of the main result of my paper in Math. Z. 118 (1970) 30–33 (mentioned at the end of the comments on p. 274).

Lecture Notes in Mathematics

A collection of informal reports and seminars
Edited by A. Dold, Heidelberg and B. Eckmann, Zürich

143

Karl W. Gruenberg
Queen Mary College, London

Cohomological Topics in Group Theory

Springer-Verlag
Berlin · Heidelberg · New York 1970

Title No. 3299

PREFACE

These notes are based on lectures that I have given at various times during the last four years and at various places, but mainly at Queen Mary College, London. Chapters 1 to 7 have been in circulation as a volume in the Queen Mary College Mathematics Notes since the autumn of 1967. They are reproduced here unchanged except for the addition of some bibliographical material and the correction of some minor errors.

Chapter 8 is an attempt at a reasonably complete survey of the subject of finite cohomological dimension. I have included proofs of everything that is not readily accessible in the literature.

Chapters 9 and 11 contain an account of a kind of globalised extension theory which I believe to be new. A survey of some of the results has appeared in volume 2 of "Category theory, homology theory and their applications", Springer Lecture Notes, no.92 (1969). The basic machinery of extension categories for arbitrary groups is given in chapter 9. Then in chapter 11 we focus attention exclusively on finite groups and primarily on the structure of minimal projective extensions. Chapter 10 is purely auxiliary and merely sets out some cohomological facts needed in chapter 11.

My aim in these lectures was to present cohomology as a tool for the study of groups. In this respect they differ basically from other available accounts of group cohomology in all of which the theory is developed with an eye on arithmetical applications. Our subject here is group theory with a cohomological flavour.

It should be stressed that there is no pretence whatsoever at completeness. In fact, the general homological machinery is kept to the bare minimum needed for the topics at hand. It follows - inevitably - that many important features are barely mentioned; and some not at all.

The audiences were not assumed to know anything about homological algebra except the most rudimentary facts. A little more knowledge of group theory was presupposed, but nothing at all sophisticated. Full references to all non-trivial or non-standard results are always given.

There is a list of the most frequently quoted books immediately following this preface. Each chapter ends with a list of all articles and books mentioned in that chapter and reference numbers refer to that list at the end of the chapter where they occur.

I was fortunate to have perceptive audiences who frequently saved me from errors and obscurities. My thanks go to all who participated and in particular to D. Cohen, I. Kaplansky, D. Knudson, A. Learner, H. Mochizuki, G. Rinehart, W. Vasconcelos and B. Wehrfritz. I owe a special debt of gratitude to Urs Stammbach for his careful and critical reading of large sections of these notes.

I am also grateful to Cornell University, the University of Oregon, the University of British Columbia and the Eidgen. Tech. Hochschule, Zürich, for financial assistance at various stages of this work.

The notes were typed by Mrs. Esther Monroe and Miss Valerie Kinsella and I thank them both for their enormous patience with me and their excellent work.

Queen Mary College,
London,
February 1970.

CONTENTS

BOOK LIST

The following books are usually referred to by their author's name only.

Burnside, W.: The theory of groups of finite order, Cambridge, 2nd edition, 1911 (Chelsea 1958).

Cartan, H. and Eilenberg, S.: Homological algebra, Princeton 1956.

Curtis, C.W. and Reiner, I.: Representation theory of finite groups and associative algebras, Interscience, 1962.

Hall, P.: Nilpotent groups, Notes of lectures at the Canadian Mathematical Congress, Univ. of Alberta, 1957. (Reprinted: Queen Mary College Mathematics Notes, 1969).

Huppert, B.: Endliche Gruppen I, Springer, 1967.

Lang, S.: Rapport sur la cohomologie des groupes, Benjamin, 1966.

Rotman, J.: The theory of groups: an introduction, Allyn and Beacon, 1965.

Schenkman, E.: Group theory, van Nostrand, 1965.

Scott, W.R.: Group theory, Prentice-Hall, 1964.

Serre, J.-P.: Corps Locaux, Hermann, 1962.

UNGEFÄHRER LEITFADEN.

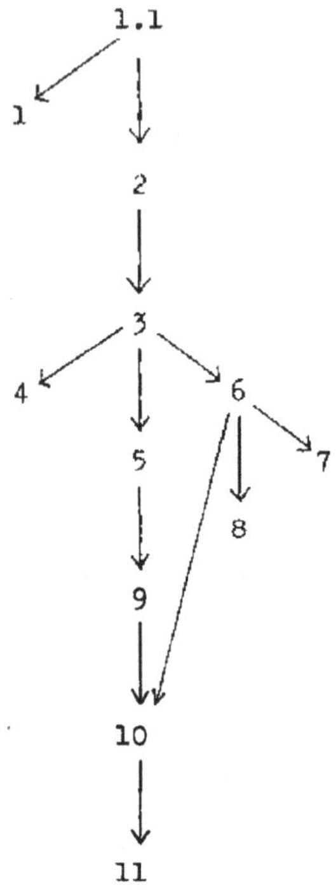

SOME NOTATION AND TERMINOLOGY

Let G be a group.

If S is a subset of a G-group M (p.1), <G<S> is the G-subgroup *generated* by S.

We write <1<S> = <S> = subgroup generated by S.

$Fr_G(M)$ = *G-Frattini group* of M (§7.1).

$d_G(M)$ = minimum number of G-generators of M (§7.1).

$N_G(A)$ = *normalizer* of A in G.

$C_G(A)$ = *centralizer* of A in G.

$|G|$, $|x|$ = order of G, x.

A complete set of representatives of the (right) cosets of A in G is called a (right) *transversal* of A in G.

G[M = M]G = *split extension* with kernel M and complement G (§1.1).

Π = product (Cartesian product).

$\coprod$ = coproduct. This is the direct sum in Mod_G, the category of G-modules. It is *, the free product, in the category of groups.

$[a,b] = a^{-1}b^{-1}ab$ = *commutator* of a by b.

If A, B are subsets of G, $[A,B] = \langle [a,b] \mid a \in A,\ b \in B \rangle$.

If H, K are subgroups of G and $K \triangleleft H$ (normal), H/K is a <u>factor</u> of G. If $[H,G] \leq K$, the factor is called <u>central</u>.

A <u>finite series</u> is a family of subgroups $(S_i;\ 0 \leq i \leq m)$, where $S_i \triangleleft S_{i+1}$.

If all factors are central, the series is called a <u>central series</u>.

If G has a finite central series from 1 to G (i.e., $S_0 = 1$ and $S_m = G$), then G is called <u>nilpotent</u>.

If G has a finite series from 1 to G with all factors abelian (cyclic), then G is called <u>soluble</u> (<u>polycyclic</u>).

h(G) = Hirsch number of the locally polycyclic G (§8.8).

If $\zeta_0(G) = 1$, $\zeta_1(G)$ = centre of G, and $\zeta_{k+1}(G)$ is the unique subgroup so that $\zeta_{k+1}(G)/\zeta_k(G) = \zeta_1(G/\zeta_k(G))$, then $(\zeta_i(G); i \geq 0)$ is called the <u>upper central series</u> of G.

If G is nilpotent and $\zeta_{c-1}(G) < \zeta_c(G) = G$, then c is the <u>class</u> of G.

If $G_1 = G$, $G_{k+1} = [G_k, G]$, then $(G_i;\ i \geq 1)$ is called the <u>lower central series</u> of G.

If $G^{(0)} = G$, $G' = [G,G]$ $(= G_2)$ and $G^{(m+1)} = (G^{(m)})'$, then $(G^{(i)};\ i \geq 0)$ is called the <u>derived series</u> of G.

CHAPTER 1

FIXED POINT FREE ACTION

1.1 The fixed point functor and its dual.

A group H is said to act on a group M if we are given a homomorphism $\theta : H \to \text{Aut } M$ (=automorphism group of M). We write $m^{h\theta}$ as m^h and call M an H-group (= H-module when M is abelian).

The element m in M is called a fixed point under H if $m^h = m$ for all h in H. We shall write the set of all these as M^H. (Warning: This notation can, on occasion, be confusing. A better one would be $C_M(H)$, the "centralizer" of H in M.)

Special case. $M \triangleleft H$ and H acts by conjugation: i.e., $m^h = h^{-1}mh$. Here M^H is all m such that mh = hm for all h in H, i.e., M^H is the centralizer of H in M in the usual sense of centralizer. In particular, M = H gives $C_H(H) = \zeta_1(H)$, the centre of H.

Remark. M^H is always a subgroup but need not be normal in M: E.g., M is the alternating group on $\{1,2,3,4\}$ and $H = \langle(1,2)\rangle$. Then $(1,2)(3,4) \in M^H$ but $(1,2)(3,4)^{(1,2,3)} = (2,3)(1,4) \notin M^H$.

M^H is the largest subgroup on which H acts trivially. There is a dual object to this: the largest image of M on which H acts trivially.

Let $[M,H] = \langle [m,h] = m^{-1}m^h;\ \text{all } m,\ h \rangle$. We assert that

(i) $[M,H]$ is normal in M

$$(\text{for } (m^{-1}m^h)^{m_1} = (mm_1)^{-1}(mm_1)^h\ (m_1{}^{-1}m_1{}^h)^{-1} \in [M,H]);$$

(ii) $[M,H]$ is H-invariant

$$(\text{for } (m^{-1}m^h)^{h_1} = (m^{-1}m^{h_1})^{-1}\ m^{-1}m^{hh_1} \in [M,H]).$$

Thus $M/[M,H]$ is an H-group on which H acts trivially and if M/K is an H-group on which H acts trivially then $K \geq [M,H]$. We write $M/[M,H] = M_H$.

The notation $[M,H]$ is meant to be reminiscent of commutators: it is in fact a group generated by commutators in the split extension E of M by H.

By definition, E is all pairs (h,m), h in H and m in M, with multiplication

$$(h,m)(h_1,m_1) = (hh_1,\ m^{h_1}m_1).$$

We shall write $E = H[M$ or $E = M]H$. We identify H and M with their images in E: i.e., h will mean $(h,1)$ and m will mean $(1,m)$. Then the given action of H on M is conjugation inside E (for $(1,m^h) = (1,m)^{(h,1)}$). Now $m^{-1}m^h = [m,h]$, the ordinary commutator; and $[M,H]$ is the subgroup generated by all these.

In verifying the normality and H-invariance of $[M,H]$, above, we really used the standard commutator formulae

$$[xy,z] = [x,z]^y[y,z],$$
$$[x,yz] = [x,z][x,y]^z.$$

<u>Special case</u>. $M \triangleleft H$. Here $M/[M,H]$ is the largest central factor with numerator M. In particular $M = H$ gives $[H,H]$, the <u>commutator group</u> of H.

1.2 Elementary consequences of fixed point free action.

Suppose M is an H-group. We say the action of H is <u>fixed point free</u> (fpf) if $M^H = 1$. This assumption can have drastic consequences for the structure of M. Here is the simplest special case:

<u>If $H = \langle\sigma\rangle$, $\sigma^2 = 1$ and M is finite, then M is commutative and of odd order.</u>

<u>Proof</u>. The mapping $x \to x^{-1} x^{\sigma}$ is one-one and therefore onto M (because M is finite!). Now $(x^{-1}x^{\sigma})^{\sigma} = (x^{-1}x^{\sigma})^{-1}$, so that $y^{\sigma} = y^{-1}$ for all y in M. Hence $y_2y_1 = (y_1{}^{-1}y_2{}^{-1})^{-1} = y_1y_2$, i.e., M is commutative. Moreover $|M|$ is odd, as y with $y^2 = 1$ implies $y^{\sigma} = y^{-1} = y$, i.e., $y = 1$.

Some condition on M is certainly necessary: e.g., if M is the free group on x,y and σ: $x \to y$, $y \to x$, then σ is fpf of order 2. (One can get away with assuming merely that each element of M has one and only one square root: B.H. Neumann [3]).

We say that an automorphism α of a group M is fpf if $m^{\alpha} \neq m$ for all $m \neq 1$, i.e., if $M^{\langle\alpha\rangle} = 1$.

If H acts on M and is cyclic with generator σ, then clearly $M^H = 1$ if, and only if, the image of σ in Aut M is a fpf automorphism. We also say σ acts fpf.

<u>Warning</u>: If H acts fpf, then the elements of H need not act fpf.

We list some simple consequences of fpf action. These are all needed in the proofs of the next section.

Let G be a group and α an automorphism of G.

(i) For any g in G, let $a = [g,\alpha]$. If α has order n, then $a\, a^{\alpha} a^{\alpha^2} \ldots a^{\alpha^{n-1}} = 1$.

(ii) If α is fpf, then $\bar{\alpha} : g \to [g,\alpha]$ is one-one.

(iii) If G is finite, α is fpf if, and only if, $\bar{\alpha}$ is surjective.

(iv) If G is finite, α is fpf and has order n, then

$$g\, g^{\alpha} \ldots g^{\alpha^{n-1}} = 1$$

for all g in G. (Apply (iii) and (i).)

(v) If G is finite, α is fpf and p is a prime dividing $|G|$, then there exists an α-invariant Sylow p-subgroup.
(If P is any p-Sylow, $P^{\alpha} = P^{x}$ for some x. Let $x = y^{-1}y^{\alpha}$ (using (iii)). Then $(P^{y^{-1}})^{\alpha} = (P^{\alpha})^{(y^{\alpha})^{-1}} = (P^{y^{-1}y^{\alpha}})^{(y^{\alpha})^{-1}} = P^{y^{-1}}$, as required.)

(vi) If G is finite, H is an α-invariant normal subgroup and α is fpf, then α is also fpf on G/H.
(Clearly $g \to [g,\alpha]$ surjective implies $gH \to [gH, \alpha]$ surjective. Now use (iii).)

(vii) Let α be fpf and $\bar{\alpha}$ be surjective. If $\bar{g}$ denotes conjugation by g, then $\alpha\bar{g} = \bar{h}^{-1}\alpha\bar{h}$ for some h in G and $\alpha\bar{g}$ is also fpf.
($g^{-1} = h^{-1}h^{\alpha}$ for some h and so, for any x in G,

$$x^{\alpha\bar{g}} = h^{-1}h^{\alpha}\, x^{\alpha}\, h^{-\alpha}h = x^{\bar{h}^{-1}\alpha\bar{h}} \quad ;$$

if x is a fixed point for $\alpha\bar{g}$, then hxh^{-1} is a fixed point for α.)

We can now prove

PROPOSITION 1. (B.H. Neumann [4].) If α is a fpf automorphism of order 3 of a group G and $\bar{\alpha}$: $g \to [g,\alpha]$ is surjective, then G is nilpotent of class ≤ 2.

Proof. By (i), $x x^{\alpha} x^{\alpha^2} = 1$ for all x in G. Take the inverse and put $y = x^{-1}$. Then $y^{\alpha^2} y^{\alpha} y = 1$ and this also holds for all y in G. Hence $[x,x^{\alpha}] = 1$ for all x. By (vii), $[x^g, x^{\alpha}] = 1$ for all g; and similarly, $[x^g, x^{\alpha^2}] = 1$. Hence $x = x^{-\alpha^2} x^{-\alpha}$ commutes with x^g, or equivalently $[g,x,x] \equiv 1$ in G. Groups with this identity and without elements of order 3 are well-known to be nilpotent of class ≤ 2. (If G had an element of order 3 then $\langle x,x^{\alpha} \rangle$ has order 9, is α-invariant and hence has a fixed point under α.)

1.3 Finite groups.

Almost all the known facts concerning fpf action deal with finite groups.

THEOREM 1. (Burnside, pp. 335-6.) Suppose M is a finite H-group and that each element of H acts fpf. Then

(i) $(|M|, |H|) = 1$;

(ii) for each p dividing $|H|$, a Sylow p-subgroup of H has exactly one subgroup of order p.

Remarks.

(1) If S is a finite p-group and there exists only one subgroup of order p, then S is cyclic or a generalized quaternion group ($= \langle a,b\colon |a| = 2^n,\ b^2 = a^{2^{n-1}},\ b^{-1}ab = a^{-1} \rangle$).

For the proof see, e.g., Scott, p. 252, or Schenkman, p. 193.

(2) A finite group H with Sylow subgroups as in Remark (1) has a known structure: if H is soluble this was determined by Zassenhaus [10]; if H is not soluble, by Suzuki [6]. (In the non-soluble case, there exists $H_1 \lhd H$ of index 1 or 2 so that $H_1 = Z \times L$, where Z is soluble with all Sylows cyclic and $L \cong SL(2,p)$.)

Proof of Theorem 1 (i). Let p divide $|H|$ and $|M|$, σ be an element of order p in H and P a σ-invariant p-Sylow of M (using (v) of §1.2). Then $\langle\sigma\rangle [P$ is a finite p-group and so there exists a non-trivial central element in P. Contradiction.

For the proof of Theorem 1 (ii) we shall need the following lemmas.

LEMMA 1. (Schenkman, p. 280.) Let M be an abelian finite group, H a subgroup of Aut M of the form $H = \langle\alpha\rangle [K$. Assume that αk is of prime order p and fpf for all k in K and that $|K|$ is prime to the exponent of M. Then K has a fixed point. In particular, if H is fpf, then K = 1.

Proof.

$$H - K = \{(\alpha k)^j,\ j = 1,\dots,p-1;\ k \in K\}:$$

For $(\alpha k)^j = (\alpha k_1)^{j_1}$ implies $j = j_1$ (by taking it modulo K); then the groups $\langle \alpha k \rangle$, $\langle \alpha k_1 \rangle$ have a non-trivial intersection, therefore are equal, and so $\alpha k = \alpha k_1$. Hence the right hand side contains exactly $|H-K|$ elements.

If $x \in M$, by (iv) of §1.2,

$$1 = \prod_{k \in K} \prod_{j=0}^{p-1} x^{(\alpha k)^j}$$

$$= x^{|K|} \prod_{j=1}^{p-1} \prod_{k \in K} x^{(\alpha k)^j}$$

$$= x^{|K|} \prod_{i=1}^{p-1} \prod_{k \in K} x^{\alpha^i k} ,$$

because also $H-K = \{\alpha^i k;\ i = 1,\dots p-1;\ k \in K\}$.

Now for each i, $\prod_{k \in K} x^{\alpha^i k}$ is a fixed point of K. As $x^{|K|} \neq 1$, there exists i so that $\prod_{k \in K} x^{\alpha^i k} \neq 1$. Qed.

LEMMA 2. If S acts on M, S is a p-group and $(|M|, p) = 1$, then for each prime q dividing $|M|$, there exists an S-invariant q-Sylow of M.

Proof. Let Q be any q-Sylow of M. If $E = S[M$, $N = N_E(Q)$ (= normalizer of Q in E), then $E = NM$ (Frattini argument). Now $|S| = |E/M| = |N/N \cap M|$ and so a p-Sylow P_1 of N is a p-Sylow of E. Hence $P_1^{\,x} = S$ for some x in E. Now Q^x is a q-Sylow whose normalizer N^x contains S .

Proof of Theorem 1 (ii). Let P be a p-Sylow of H. It is sufficient to prove that every subgroup P of order p^2 must be cyclic. So let $S = \langle s\rangle \times \langle s'\rangle$ be of order p^2. By Theorem 1 (i), $p \nmid |M|$. Take a prime $q/|M|$ and a q-Sylow Q of M invariant under S (Lemma 2). Let A be the centre of Q. Apply Lemma 1 to A (for M) and S(for H) and get $\langle s'\rangle = 1$, a contradiction.

We turn now to consequences to the structure of M.

THEOREM 2. *(J. Thompson [7].) If M is a finite group with a fpf automorphism α of prime order p, then M is nilpotent.*

As a first step we prove the following special case.

LEMMA 3. *(Cf. G. Higman [2]). Let α be a fpf automorphism of prime order p of the finite soluble group M. Then M is nilpotent.*

Proof. Use induction on $|M|$. It will be sufficient to prove $\zeta_1(M) = Z > 1$. For Z is certainly α-invariant and so α acts fpf on M/Z ((vi), §1.2). By induction, M/Z is nilpotent and therefore M is nilpotent.

Let Q be a minimal α-invariant normal subgroup of M. (Form $E = \langle\alpha\rangle [M$ and take Q as a minimal normal subgroup of E contained in M.) So Q is an elementary q-group for some prime q and $q \neq p$ by Theorem 1 (i). If M is not a q-group, choose a prime $r \neq q$ and an α-invariant r-Sylow R ((v), §1.2). If $QR < M$, then QR is nilpotent (induction) and so $R \leq C(Q)$. Thus either $M = QR$ for some r or $(M:C(Q))$ is a q-power: but then $Z > 1$.

So assume $M = QR$, let K be the image of R in Aut Q via conjugation and form $H = \langle\alpha\rangle [K$. By (vii) of §1.2, αk is of order p, for all k in K, and acts fpf on Q. Now we can apply Lemma 1 to Q, H and conclude K has a fixed point, or equivalently, M has a centre.

Theorem 2 depends on the following result also due to Thompson [8].

<u>THOMPSON'S p-COMPLEMENT THEOREM.</u> <u>Let p be an odd prime dividing $|G|$, P a Sylow p-subgroup of G and Z the centre of P. Define $J = J(P)$ to be the subgroup generated by all abelian subgroups of P for which the minimum number of generators is the maximum possible. If $C_G(Z)$ and $N_G(J)$ have normal p-complements, then so does G.</u>

<u>Proof of Theorem 2</u>. By Lemma 3 we need only show M is soluble. If M is a 2-group, we are finished. If not, choose $q \neq 2$, $q/|M|$ and find an α-invariant q-Sylow Q of M. Let $Z = \zeta_1(Q)$, $J = J(Q)$. Both are characteristic subgroups in Q and therefore α-invariant. If Z (or J) is normal in M, then α induces a fpf automorphism on M/Z (or M/J). So by induction, M/Z (or M/J) is nilpotent and therefore M is soluble. Hence we may assume $N(J) < M$ and $N(Z) < M$. By induction, both $C(Z)$, $N(J)$ are nilpotent and so have normal q-complements. By Thompson's theorem above, M has a normal q-complement, say R. Now R is characteristic and so α-invariant. By induction, R is nilpotent and again M is soluble.

Theorem 2 does not completely settle the structure of groups having a fpf automorphism of prime order: what nilpotent groups can arise? In this direction we have a result of Graham Higman [2]: there exists a function $f : \{\text{primes}\} \to \mathbb{Z}_{>0}$ so that if M is any nilpotent group (not necessarily finite) with a fpf automorphism of order p, then class $M \leq f(p)$.

If H is not of prime order, our knowledge becomes meagre. Let $\Delta(H)$ denote the class of all finite groups of orders prime to $|H|$ on which H can act fpf.

(1) All the groups in $\Delta(H)$ are nilpotent if, and only if, H has prime order.

(The "if" part is Theorem 2; the "only if" part is essentially in Shult [5].)

(2) All the groups in $\Delta(H)$ have nilpotent commutator groups if, and only if, H is of order 4 or is the symmetric group of degree 3. (The "if" part in case H is cyclic of order 4 is in Gorenstein-Herstein [1]; when H is the Klein group it is due to S. Bauman [11]; the rest is all in Shult [5].)

To conclude we mention one important general fact, due to Thompson [9].

Define, for any group G, F(G) to be the product of all nilpotent normal subgroups. (This is nilpotent if G is finite.) It is called the Fitting group of G. The upper Fitting series is then

$$1 \leq F_1 \leq F_2 \leq \dots ,$$

where $F_{i+1}/F_i = F(G/F_i)$.

If G is soluble, there exists a first m so that $F_m = G$. Call m the Fitting height of G and write $m = f(G)$.

THEOREM. (J. Thompson [9].) Let H,M be finite soluble groups of coprime orders. If $|H| = p_1 \dots p_n$ (where p_i are primes but not necessarily distinct), then

$$f(M) \leq 5^n f(M^H).$$

So in particular, if $M^H = 1$, $f(M) \leq 5^n$.

Note: the bound is quite extravagant: the case $|H| = 4$ and $M^H = 1$ gives $f(M) \leq 25$, while actually $f(M) \leq 2$.

Sources and References

Sections 1.2 and 1.3 are mainly based on the last section of Schenkman's book.

[1] Gorenstein, D. and Herstein, I.N. : Finite groups admitting a fixed point free automorphism of order 4, Amer. J. Math. 83 (1961) 71-78.

[2] Higman, G.: Groups and rings having automorphisms without non-trivial fixed elements, J. London Math. Soc. 32 (1957) 321-334.

[3] Neumann, B.H.: On the commutativity of addition, J. London Math. Soc., 15 (1940) 203-208.

[4] Neumann, B.H.: Groups with automorphisms that leave only the neutral element fixed, Archiv d Math. 7 (1956) 1-5.

[5] Shult, E.: Nilpotence of the commutator subgroup in groups admitting fixed point free operator groups, Pacific J. Math. 17 (1966) 323-347.

[6] Suzuki, M.: On finite groups with cyclic Sylow subgroups for all odd primes, Amer. J. Math. 77 (1955) 657-691.

[7] Thompson, J.: Finite groups with fixed point free automorphisms of prime order, Proc. Nat. Acad. Sci. U.S.A. 45 (1959) 578-581.

[8] Thompson, J.: Normal p-complements for finite groups, J. of Algebra 1 (1964) 43-46.

[9] Thompson, J.: Automorphisms of solvable groups, J. of Algebra 1 (1964) 259-267.

[10] Zassenhaus, H.: Über endliche Fastkörper, Abh.math.Sem. Hamburg, 11 (1936) 187-220.

[11] Bauman, S.F.: The Klein group as an automorphism group without fixed points, Pacific J. of Math. 18 (1966) 9-13.

[7] contains an excellent historical survey. [2] contains a wealth of interesting material not alluded to above - in particular, a solution of an analogous problem for Lie rings.

The material on f.p.f. automorphisms can now be read in full detail in Daniel Gorenstein's book Finite Groups (Harper and Row, 1968).

CHAPTER 2

THE COHOMOLOGY AND HOMOLOGY GROUPS

2.1 The cohomology functor.

The next simplest case to having no fixed points at all is when the fixed points lie in an abelian group. The study of this case leads to the cohomology theory of groups.

We consider a fixed group G and the category of abelian G-groups, i.e., of G-modules. (If A is a G-module we shall write ag instead of a^g. All modules will be right modules except when stated otherwise.) The covariant functor $A \mapsto A^G$ is left exact but is not in general exact: e.g., consider

$$0 \to \mathbb{Z} \to \mathbb{Z} \to \mathbb{Z}/2\mathbb{Z} \to 0$$

and let $G = \mathrm{Aut}\,\mathbb{Z}$.

Any G-module can be viewed as a module over $\mathbb{Z}G$ (= group ring of G) and vice versa.

If B is a left module and A is a right module we can make $H = \mathrm{Hom}_{\mathbb{Z}}(B,A)$ into a right G-module by setting

$$b(f^x) = (xb)f\,x,$$

where $f \in H$, $x \in G$, $b \in B$.

Any left module B can be made into a right module by setting $bx = x^{-1}b$ (and vice versa). Note that then $H^G = \mathrm{Hom}_{\mathbb{Z}G}(B,A)$.

<u>Warning</u>: If B is a two-sided module we can form H as above using the left module structure on B. But we can also use the right module structure on B to make B into a new left module and then get a new G-module structure on H - call this H'. We need <u>not</u> have $H \cong H'$. (Example: Suppose the right structure on B is non-trivial but that the left structure is trivial and compare H^G, $(H')^G$, with A trivial and A = B as additive group.)

<u>Exercise</u>. If H, H' are as in the last paragraph, but with $B = \mathbb{Z}G$, then $H \cong H'$. (If ι is the $\mathbb{Z}$-automorphism $x \to x^{-1}$, $f \to \iota f$ is a possible G-isomorphism.)

<u>Definition</u>. If C is an additive group, put $C^* = \mathrm{Hom}_{\mathbb{Z}}(\mathbb{Z}G, C)$ and view C^* as a right G-module using the left structure on $\mathbb{Z}G$ and the trivial structure on C. (Thus $uf^x = xuf$.) A G-module of the form C^* is called a <u>coinduced module</u>.

If A is a G-module we form A^* by first forgetting the module structure on A. Then $a \to a^*$, where $a^*: x \to ax$, is a G-monomorphism $A \to A^*$.

Note that $C \mapsto C^*$ is an exact functor (from the category of abelian groups to that of G-modules).

<u>Definition</u>. A <u>cohomological extension</u> of the functor $A \mapsto A^G$ is a sequence of functors $H^q(G, \ )$, $q = 0,1,2,\ldots$, (from G-modules to abelian groups) so that

$$H^0(G,A) = A^G ,$$

together with a connecting homomorphism $d = (d^0, d^1, \ldots)$, where

$$d^q\colon H^q(G,C) \to H^{q+1}(G,A)$$

whenever $0 \to A \to B \to C \to 0$ is a given exact sequence of G-modules. It is assumed further that the resulting "cohomology sequence" is exact; and that d is natural: i.e., if

$$\begin{array}{ccccccccc} 0 & \to & A & \to & B & \to & C & \to & 0 \\ & & \downarrow & & \downarrow & & \downarrow & & \\ 0 & \to & A' & \to & B' & \to & C' & \to & 0 \end{array}$$

is a given commutative diagram with exact rows, then the resulting diagram involving the two cohomology sequences is also commutative.

The cohomological extension $(H^q(G,\ ))$ is called minimal if $H^q(G,A) = 0$ for all $q > 0$ and all coinduced A.

Definition. If $((\overline{H}^q(G,\ )), (\overline{d}^q))$ is a second minimal cohomological extension then we say $(H^q(G,\ )), (\overline{H}^q(G,\ ))$ are naturally isomorphic (naturally equivalent) if, for each A and q, we are given an isomorphism

$$\phi_A^q : H^q(G,A) \cong \overline{H}^q(G,A)$$

such that $A \to B$ always yields a commutative square

$$\begin{array}{ccc} H^q(G,A) & \cong & \overline{H}^q(G,A) \\ \downarrow & & \downarrow \\ H^q(G,B) & \cong & \overline{H}^q(G,B) \end{array}$$

and the connecting homomorphisms commute appropriately: i.e., if $0 \to A \to B \to C \to 0$ is exact, then

$$\begin{array}{c}\text{cohomology sequence for } H\\ \downarrow\\ \text{cohomology sequence for } \overline{H}\end{array}$$

commutes.

THEOREM 1. There always exists a minimal cohomological extension of $A \mapsto A^G$ and any two such are naturally isomorphic.

Proof.

Uniqueness. We postpone proving the existence of a natural isomorphism between two extensions: this is a special case of Theorem 1 of Chapter 6 (§6.1). We merely make here the simple observation that two extensions give naturally isomorphic functors at each dimension $q \geq 0$.

For if A is given, we embed A in A^*: $0 \to A \to A^* \to B \to 0$. Then the corresponding exact cohomology sequence

$$0 \to A^G \to (A^*)^G \to B^G \to H^1(G,A) \to 0 \to H^1(G,B) \to H^2(G,A) \to 0 \to \dots$$

shows that H^1 is unique (to within an isomorphism); then that H^2 is unique; and so on.

Existence. Take a projective resolution of $\mathbb{Z}$:

$$\dots \to P_2 \to P_1 \to P_0 \to \mathbb{Z} \to 0.$$

Form $(\mathrm{Hom}_{\mathbb{Z}G}(P_i,A))$ and take the homology of this complex. We verify that this functor, call it $(H^q(A))$, is a minimal cohomological extension.

(i) $H^0(A) = \mathrm{Hom}_G(\mathbb{Z},A) \cong A^G$.

(ii) Let A be coinduced, i.e., $A = C^*$ for some additive group C. So

$$\mathrm{Hom}_G(P_i,A) \cong \mathrm{Hom}_{\mathbf{Z}}(P_i,C) = K^i \text{ , say.}$$

Now $0 \to A^G \to K^0 \to K^1 \to \ldots$ is exact (because the original resolution splits over $\mathbf{Z}$) and therefore its homology is zero.

(iii) Connecting homomorphisms.

As P_i is projective, the following sequence is exact:

$$0 \to \mathrm{Hom}_G(P_i,A) \to \mathrm{Hom}_G(P_i,B) \to \mathrm{Hom}_G(P_i,C) \to 0.$$

Write

$$L_{i,A} = \mathrm{Hom}_G(P_i,A), \text{ etc.,}$$

$$K_{i,A} = \mathrm{Ker}\ (L_{i,A} \to L_{i+1,A}), \text{ etc.,}$$

$$\overline{K}_{i,A} = L_{i,A}/\ \mathrm{Im}(L_{i-1,A} \to L_{i,A}), \text{ etc..}$$

It is trivial to check that

$$0 \to K_{i,A} \to K_{i,B} \to K_{i,C}$$

is exact and also

$$\overline{K}_{i,A} \to \overline{K}_{i,B} \to \overline{K}_{i,C} \to 0.$$

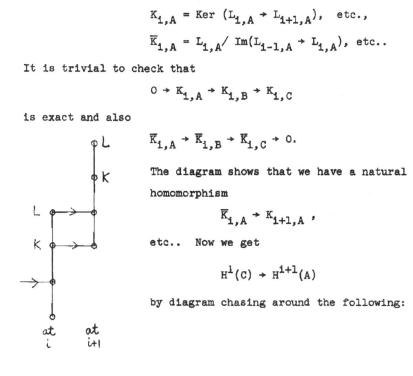

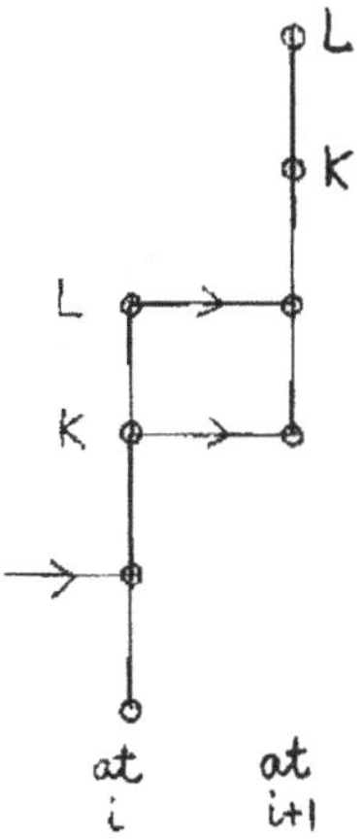

The diagram shows that we have a natural homomorphism

$$\overline{K}_{i,A} \to K_{i+1,A} \text{ ,}$$

etc.. Now we get

$$H^i(C) \to H^{i+1}(A)$$

by diagram chasing around the following:

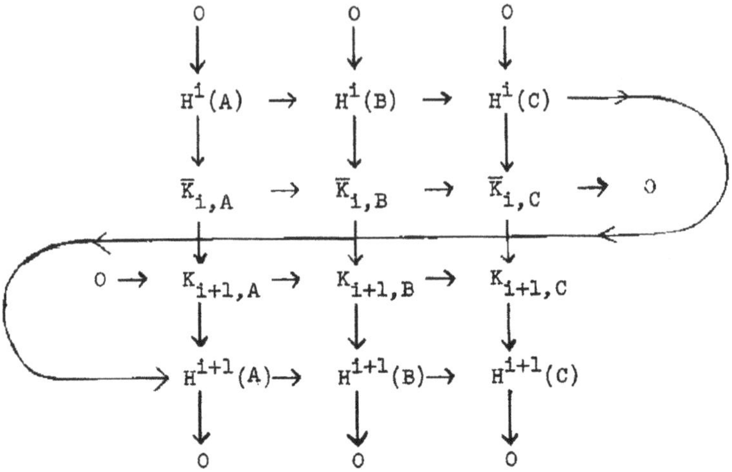

For details of the exactness of the cohomology sequence and the naturality of d, cf. Bourbaki, Algèbre Commutative, chapter 1 (or, of course, Cartan-Eilenberg, chapter 3).

Definition. If $(H^q(G,\ );\ d)$ is a minimal cohomological extension of $A \mapsto A^G$, we call $H^q(G,A)$ the *q-dimensional* (or q-th) *cohomology group of G with coefficients in A*.

PROPOSITION 1. Let $\cdots \to P_2 \to P_1 \to P_0 \to \mathbb{Z} \to 0$ *be a projective resolution of* $\mathbb{Z}$ *and* $Y = \operatorname{Im}(P_q \to P_{q-1})$*. Then for any G-module A,*

$$\operatorname{Hom}_G(P_{q-1}, A) \to \operatorname{Hom}_G(Y, A) \to H^q(G,A) \to 0$$

is exact.

Proof.

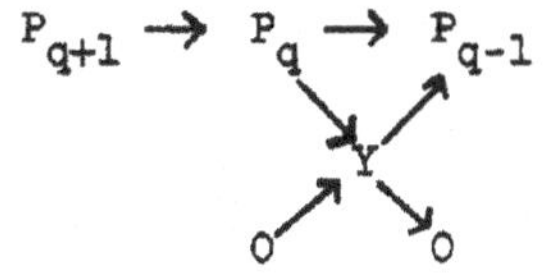

$$P_{q+1} \to P_q \to P_{q-1}$$
$$\nearrow Y \searrow$$
$$0 \qquad 0$$

yields

$$\text{Hom}_G(P_{q+1},A) \leftarrow \text{Hom}_G(P_q,A) \leftarrow \text{Hom}_G(P_{q-1},A) \;.$$
$$\text{Hom}_G(Y,A)$$
$$0$$

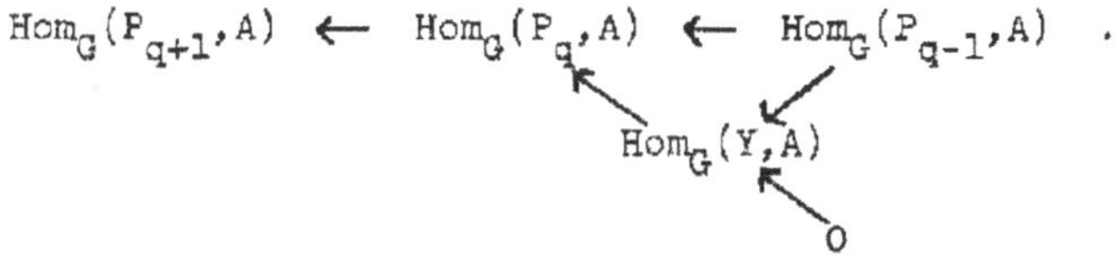

Now apply the definition of $H^q(G,A)$ as the q-th homology group
of the complex $(\text{Hom}_G(P_1,A))$.

2.2 The homology functor.

If A is a G-module, then in the usual additive notation,
$[a,g] = -a + ag = a(g-1)$ (cf. §1.1). Hence $[A,G]$ is the additive
subgroup generated by all $a(g-1)$. This is a submodule (cf. the
general case: $[A,G]$ is always G-invariant).

If $\mathfrak{g}$ is the additive subgroup of $\mathbb{Z}G$ generated by all $g-1$,
$g \in G$, then $[A,G] = A\,\mathfrak{g}$. Thus $A_G = A/A\,\mathfrak{g}$. The mapping $G \to 1$
extends by $\mathbb{Z}$-linearity to a ring homomorphism ϵ of $\mathbb{Z}G$ onto $\mathbb{Z}$ and
clearly $\mathfrak{g}$ is the kernel of ϵ. We call ϵ the (unit) augmentation
of $\mathbb{Z}G$ and $\mathfrak{g}$ the augmentation ideal of G (or of $\mathbb{Z}G$). The pair
$(\mathbb{Z}G, \epsilon)$ is an example of an augmented ring.

If C is an additive group, we make $C_* = C \otimes_{\mathbb{Z}} \mathbb{Z}G$ into a right G-module using the right structure on $\mathbb{Z}G$. Any module of this form is called an <u>induced module</u>. (Example: free modules.)

<u>Remark. If G is finite, induced = coinduced.</u>

<u>Proof</u>. We map $C \otimes_{\mathbb{Z}} \mathbb{Z}G \to \mathrm{Hom}_{\mathbb{Z}}(\mathbb{Z}G, C)$ as follows: given c in C, g in G, let c.g be the $\mathbb{Z}$-linear extension of the mapping

$$u \to \begin{cases} c & \text{if } ug = 1 \\ 0 & \text{if } ug \neq 1 \end{cases} .$$

Then $\cdot$ is bilinear and so yields a homomorphism $C_* \to C^*$. This is easily checked to be a G-monomorphism. The finiteness of G is invoked only now to ensure that the mapping is surjective.

It follows that for a <u>finite</u> group G, we have $H^q(G,A) = 0$ for all $q > 0$ whenever A is a free $\mathbb{Z}G$-module and therefore also when A is projective.

<u>Definition</u>. A <u>minimal homological extension</u> of $A \mapsto A_G$ is a sequence of functors $H_q(G, \ )$, $q = 0,1,2,\ldots$, such that

$$H_0(G,A) = A_G \quad ,$$

together with a connecting homomorphism $d = (d_0, d_1, \ldots)$, where

$$d_q\colon H_{q+1}(G,C) \to H_q(G,A)$$

whenever $0 \to A \to B \to C \to 0$ is a given exact sequence of G-modules. The resulting "homology sequence" is assumed to be exact and d is to be natural. Moreover, $H_q(G,A) = 0$ for all $q > 0$ whenever A is induced.

A <u>natural isomorphism</u> between two such extensions (H_q), $(\bar{H}_q)$ is defined in the expected way: cf. the corresponding notion in cohomology.

<u>THEOREM 2. There always exists a minimal homological extension of</u> $A \mapsto A_G$ <u>and any two such are naturally isomorphic</u>.

<u>Sketch of proof</u>.

<u>Uniqueness</u>. Take any G-module A, view it as an additive group only, form A_* and project $A_* \to A$ by $a \otimes g \to ag$. Then the exact sequence $0 \to B \to A_* \to A \to 0$ gives

$$\ldots \to 0 \to H_2(G,A) \to H_1(G,B) \to 0 \to H_1(G,A) \to H_0(G,B) \to H_0(G,A_*) \to H_0(G,A) \to 0$$

This shows that H_1 is unique; then H_2 is unique; etc..

<u>Existence</u>. Take a <u>left</u> projective resolution of $\mathbb{Z}$ (i.e., each P_i is a left projective G-module):

$$\ldots \to P_2 \to P_1 \to P_0 \to \mathbb{Z} \to 0.$$

Form the complex

$$\ldots \to A \otimes_G P_1 \to A \otimes_G P_0 \to A \otimes_G \mathbb{Z} \to 0$$

and take its homology.

One checks that this is a possible homological extension in much the same way as in the cohomology case. Note that $H_0(G,A) = A_G$ because $0 \to \mathfrak{I} \to \mathbb{Z}G \to \mathbb{Z} \to 0$ when tensored with A on the left gives $A \otimes_G \mathbb{Z} \cong A/\mathrm{Im}(A \otimes \mathfrak{I}) = A/A\mathfrak{I}$. In the connecting homomorphism argument, the exactness of $\mathrm{Hom}_G(P_i, \ )$ is here replaced by the exactness of $\otimes_{\mathbb{Z}G} P_i$.

<u>Definition</u>. If $(H_q(G,\ );\ d)$ is a minimal homological extension of $A \mapsto A_G$, we call $H_q(G,A)$ the <u>q-dimensional</u> (or q-th) <u>homology group of G with coefficients in A</u>.

<u>Remark</u>. If one works with a left module A one must use a right resolution. But all essentials are unchanged.

<u>Note</u>: Let A be a right G-module and A' denote A when made into a left module (i.e., $g.a = ag^{-1}$). Suppose P is a left projective resolution of $\mathbb{Z}$ and P' is the corresponding right one. Then

$$\text{Homology of } (A \underset{\mathbb{Z}G}{\otimes} P) \cong \text{Homology of } (P' \underset{\mathbb{Z}G}{\otimes} A')$$

because $A \otimes P \cong P' \otimes A'$ via $a \otimes x \leftrightarrow x \otimes a$. Hence there are natural isomorphisms

$$H_q(G,A) \cong H_q(G,A').$$

<u>PROPOSITION 2. If</u> $\cdots P_2 \to P_1 \to P_0 \to \mathbb{Z} \to 0$ <u>is a right projective resolution of $\mathbb{Z}$ and</u> $Y = \mathrm{Im}(P_q \to P_{q-1})$, <u>then, for any left G-module A,</u>

$$0 \to H_q(G,A) \to Y \underset{G}{\otimes} A \to P_{q-1} \underset{G}{\otimes} A$$

<u>is exact</u>.

<u>Proof</u> = exercise. (Cf. Proposition 1.)

2.3 Change of coefficient ring.

Modules frequently come to us endowed with additional structure. A typical situation occurs when $A \triangleleft G$ and A is an elementary abelian p-group. Then A is a G-module (via conjugation) and also a vector space over $\mathbb{F}_p$ (=field of p elements). So A is really a module over $\mathbb{F}_p G$. How does this affect the cohomology and homology of G with coefficients in A? Answer: not at all; as we now show.

Let R be any commutative ring and form RG (= the group algebra over R). We may repeat all the definitions and constructions of §2.1 and §2.2.

Any RG-module is, in particular, a G-module and therefore a $\mathbb{Z}G$-module. Conversely, for any $\mathbb{Z}G$-module A,

$$_{(R)}A = R \otimes_{\mathbb{Z}} A$$

is a module over RG. Moreover, A is $\mathbb{Z}G$-projective implies $_{(R)}A$ is RG-projective. (Use: projective = direct summand of free.)

If $\ldots \to P_1 \to P_0 \to \mathbb{Z} \to 0$ is a $\mathbb{Z}G$-projective resolution of $\mathbb{Z}$, applying $_{(R)}\cdot$ gives an RG-projective resolution of R. Now for any <u>RG-module</u> A, there is a natural isomorphism

$$\mathrm{Hom}_R({}_{(R)}P, A) \cong \mathrm{Hom}_{\mathbb{Z}}(P,A) \ ;$$

and this is automatically a G-module isomorphism. Take invariant elements and we get

$$\mathrm{Hom}_{RG}({}_{(R)}P, A) \cong \mathrm{Hom}_{\mathbb{Z}G}(P, A).$$

So the homology of the left hand complex coincides with that of the right hand one; i.e., with $H^*(G,A)$.

<u>Conclusion</u>: <u>We may also calculate $H^q(G,A)$ by taking any RG-projective resolution of R.</u>

<u>A similar conclusion for $H_q(G,A)$.</u>

Remark. The above really shows that for all RG-modules A,

$$\mathrm{Ext}^*_{RG}(R,A) \cong \mathrm{Ext}^*_{\mathbb{Z}G}(\mathbb{Z},A) ;$$

and

$$\mathrm{Tor}^{RG}_*(R,A) \cong \mathrm{Tor}^{\mathbb{Z}G}_*(\mathbb{Z},A).$$

2.4 Isomorphism of group rings.

Suppose G, H are groups and that there exists a ring isomorphism $\alpha : \mathbb{Z}G \to \mathbb{Z}H$ so that

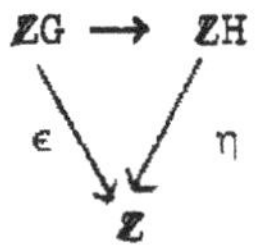

commutes, where ϵ, η are the respective unit augmentations. We then say that α is an isomorphism of augmented rings:

$$\alpha : (\mathbb{Z}G, \epsilon) \to (\mathbb{Z}H, \eta).$$

Quite generally, whenever R, R^* are rings and there is a homomorphism $\mu : R \to R^*$, we obtain a functor from R^*-modules to R-modules: if A is an R^*-module, then the additive group of A is made an R-module, call it $A_{(\mu)}$, by $ar = a\, r^{\mu}$ (r in R). (Exercise: Write down the effect of the functor on morphisms.)

Returning to α: $(\mathbb{Z}G, \epsilon) \to (\mathbb{Z}H, \eta)$, we see that $\mathbb{Z}$ a trivial H-module also implies $\mathbb{Z}_{(\alpha)}$ is trivial for G. Hence G and H have the same cohomology (i.e., $H^q(G, A_{(\alpha)}) \cong H^q(H,A)$ for all q and all A) and also the same homology (i.e., $H_q(G, A_{(\alpha)}) \cong H_q(H,A)$ for all q, A). (The isomorphisms are all natural. More precisely: the functors $(H^q(G, \cdot_{(\alpha)}), d)$, $(H^q(H, \ ), d)$ are isomorphic; and similarly for homology.)

PROPOSITION 3. (Cf. for example: Cartan-Eilenberg, p. 183.) If G, H are groups and there exists a ring isomorphism $\mathbb{Z}G \xrightarrow{\sim} \mathbb{Z}H$, then there exists an augmented ring isomorphism $(\mathbb{Z}G, \epsilon) \xrightarrow{\sim} (\mathbb{Z}H, \eta)$. Hence G and H have the same cohomology and the same homology.

Proof. Given $\beta : \mathbb{Z}G \xrightarrow{\sim} \mathbb{Z}H$. So $\gamma = \beta\eta : \mathbb{Z}G \to \mathbb{Z} \to 0$. Now as g is a unit in $\mathbb{Z}G$, $g\gamma$ is one in $\mathbb{Z}$ and so $g\gamma = \pm 1$. Let α be the $\mathbb{Z}$-linear extension of the mapping $G \to \mathbb{Z}G$ defined by $g\alpha = (g\gamma)g$. Then

$$(xy)\alpha = (xy)\gamma\ (xy) = x\gamma\ y\gamma\ xy = x\alpha\ y\alpha,$$

and so α is a ring automorphism of $\mathbb{Z}G$ and $g\alpha\gamma = 1$ for all g, i.e., $\alpha\gamma$ is the unit augmentation. Now clearly

$$\begin{array}{ccc} \mathbb{Z}G & \xrightarrow{\alpha\beta} & \mathbb{Z}H \\ & {\scriptstyle\alpha\gamma}\searrow \quad \swarrow{\scriptstyle\eta} & \\ & \mathbb{Z} & \end{array}$$

commutes, as required.

The following problem is still wide open: What does $\mathbb{Z}G \cong \mathbb{Z}H$ imply about the relation between G and H? Does it perhaps even force G and H to be isomorphic?

A little progress has been made with this problem in the case of finite groups. If $\mathbb{Z}G \cong \mathbb{Z}H$ then for any field K, $KG \cong KH$ and if our groups are finite then their representation theory over K can be invoked to yield some information. Cf. [1] and the literature cited there.

We make one general remark here. By Proposition 3, $\mathbb{Z}G \cong \mathbb{Z}H$ implies that we have an augmented ring isomorphism and so $\mathfrak{g} \cong \mathfrak{h}$ (augmentation ideals). Now

$$\mathfrak{g}/\mathfrak{g}^2 \cong G/G' :$$

For $\mathfrak{g}$ is the free abelian group on all 1-a and hence $1-a \to aG'$ extends to a homomorphism $\mathfrak{g} \to G/G'$ whose hernel contains $\mathfrak{g}^2$ (because $(1-a)(1-b) = (1-a) + (1-b) - (1-ab)$). The resulting homomorphism $\mathfrak{g}/\mathfrak{g}^2 \to G/G'$ has inverse $aG' \to 1-a + \mathfrak{g}^2$. Thus we have shown that $\mathbb{Z}G \cong \mathbb{Z}H$ implies $G/G' \cong H/H'$.

Sources and references.

Our introduction of cohomology and homology is in the spirit of Serre's Corps Locaux, chapter 7. For §2.3 confer Cartan-Eilenberg, chapter 10, §2.

[1] Obayashi, T.: Solvable groups with isomorphic group algebras, J. Math. Soc. Japan, 18 (1966) 394-397.

CHAPTER 3

PRESENTATIONS AND RESOLUTIONS

3.1 A functor from presentations to resolutions

Let G be a group and $1 \to R \to F \overset{\pi}{\to} G \to 1$ any presentation with F free.

The augmentation ideals of F, G will be denoted by $\mathfrak{f}, \mathfrak{g}$ respectively; and we write $\mathfrak{r} = \text{Ker } (\mathbb{Z}F \to \mathbb{Z}G)$. All these are two-sided ideals.

Our aim is to show that the above presentation leads in a natural way to a free G-resolution of $\mathbb{Z}$.

PROPOSITION 1. If F is free on a set X, $\mathfrak{f}$ is free, as right $\mathbb{Z}F$-module, on 1-X.

Clearly 1-X generates $\mathfrak{f}$ as right ideal - for the right ideal generated by 1-X lies in $\mathfrak{f}$ and $\mathbb{Z}F$ is trivial modulo this ideal, so that it contains $\mathfrak{f}$.

To prove freeness, we need: every mapping of 1-X into a module extends to a module homomorphism of $\mathfrak{f}$.

Translation to group theory: first step.

Definition. A mapping d of a group G into a G-module M is called a derivation (or crossed homomorphism) if $(xy)d = (xd)y + yd$ for all x, y in G.

LEMMA 1. *Let $d : G \to M$ be a mapping and define $\delta : \mathfrak{g} \to M$ by $(1-g)\delta = gd$. Then d is a derivation if, and only if, δ is a module homomorphism.*

Proof. $((1-x)y)\delta = (xy)d - yd$ and the left hand side is equal to $(1-x)\delta y = (xd)y$ if, and only if, d is a derivation.

So Proposition 1 is reduced to proving: any mapping of X into an F-module extends to a derivation of F.

Translation to group theory: second step.

The following result is clear.

LEMMA 2. *Let G be any group, M a G-module and $E = G[M$ (split extension). Let $s : G \to E$ be a mapping such that $gs = (g, gd)$. Then s is a homomorphism if, and only if, d is a derivation.*

Thus Proposition 1 is reduced to proving: if M is any F-module and E is the split extension of M by F, then any mapping $x \to (x, x')$, $x \in X$, extends to a homomorphism : $F \to E$. But this is clear. So Proposition 1 is proved.

As a corollary of Proposition 1 we have the following result which contains the proposition (the case $R = F$).

THEOREM 1. *If $R \lhd F$ and R is free on a set Y, then $\mathfrak{r} = \mathrm{Ker}(\mathbb{Z}F \to \mathbb{Z}(F/R))$ is free as right $\mathbb{Z}F$-module on $1-Y$.*

Proof. Recall that, by Schreier's theorem (e.g. Rotman, p. 242), R *is* free and hence Y does always exist.

Let $\mathfrak{r}'$ be the right ideal generated by 1-Y. Obviously, $\mathfrak{r}' \leq \mathfrak{r}$. Also, $\mathfrak{r}'$ contains the augmentation ideal of R (Proposition 1, generation part) and hence $\mathfrak{r}'$ is the right ideal generated by all 1-w, w$\in$R. Now $\mathbb{Z}F/\mathfrak{r}'$ is an F/R-module and so $\mathfrak{r} \leq \mathfrak{r}'$. Hence $\mathfrak{r}$ is generated by 1-Y.

Suppose $\Sigma\,(1-y_i)\alpha_i = 0$, $\alpha_i \in \mathbb{Z}F$. Let T be a transversal (= complete set of coset representatives) of R in F; and $\alpha_i = \Sigma\,\beta_{ij}t_j$ with $\beta_{ij} \in \mathbb{Z}R$. So our relation implies the coefficient of each t_j is 0: i.e., $\sum_i (1-y_i)\beta_{ij} = 0$, all j. But this implies that each $\beta_{ij} = 0$ (Proposition 1, freeness).

Remark. $\mathfrak{r}$ is actually a two-sided ideal. It is also free on 1-Y when viewed as left F-module. (Repeat the above proof.)

LEMMA 3. If $\mathfrak{a}$ is a right ideal of $\mathbb{Z}F$, $\mathfrak{a}/\mathfrak{a}\mathfrak{r}$ is a right G-module.

Proof. If $\alpha \in \mathfrak{a}$ and $g = w\pi \in G$, then $(\alpha + \mathfrak{a}\mathfrak{r})g = \alpha w + \mathfrak{a}\mathfrak{r}$.

LEMMA 4. (i) If $\mathfrak{a}$ is free as right ideal of $\mathbb{Z}F$ on S, $\mathfrak{a}/\mathfrak{a}\mathfrak{r}$ is G-free on S + $\mathfrak{a}\mathfrak{r}$.

(ii) If $\mathfrak{a}$ is free as right ideal on S, $\mathfrak{b}$ is free as right ideal on T and is also two-sided, then $\mathfrak{a}\mathfrak{b}$ is free as right ideal on ST.

(Ditto with right and left interchanged.)

Proof. (i) $\mathfrak{a} = \coprod_{s\in S} s(\mathbb{Z}F)$; $\mathfrak{a}\mathfrak{r} = \coprod_{s\in S} s\mathfrak{r}$ and so $\mathfrak{a}/\mathfrak{a}\mathfrak{r} \cong \coprod_{s\in S} s(\mathbb{Z}G)$.

(ii) is clear.

Theorem 1 and Lemmas 3, 4 now combine to give

THEOREM 2. (Gruenberg [1].) *(i) If*

$$1 \to R \to F \xrightarrow{\pi} G \to 1\ , \tag{P}$$

with F a free group, is exact, then the following is a free G-resolution of $\mathbb{Z}$:

$$\ldots \to \mathfrak{r}^2/\mathfrak{r}^3 \to \mathfrak{f}\mathfrak{r}/\mathfrak{f}\mathfrak{r}^2 \to \mathfrak{r}/\mathfrak{r}^2 \to \mathfrak{f}/\mathfrak{f}\mathfrak{r} \to \mathbb{Z}G \to \mathbb{Z} \to 0\ , \tag{R}$$

where $\mathbb{Z}G \to \mathbb{Z}$ *is the unit augmentation,* $\mathfrak{f}/\mathfrak{f}\mathfrak{r} \to \mathbb{Z}G$ *is induced by* π *and the remaining mappings are all induced by the appropriate inclusions.*

(ii) If F is free on X and R is free on Y, then $\mathfrak{r}^n/\mathfrak{r}^{n+1}$ *is G-free on the cosets of all elements*

$$(1-y_1)\ldots(1-y_n), \quad y_1,\ldots,y_n \text{ in } Y;$$

and $\mathfrak{f}\mathfrak{r}^{n-1}/\mathfrak{f}\mathfrak{r}^n$ *is G-free on the cosets of all elements*

$$(1-x)(1-y_1)\ldots(1-y_{n-1}),\ x \text{ in } X, \quad y_1,\ldots,y_{n-1} \text{ in } Y.$$

(iii) (P) $\mapsto$ (R) is a covariant functor from the category of presentations of G under free groups to the category of free G-resolutions of $\mathbb{Z}$.

Exercise: What precisely are the morphisms in the categories of part (iii)?

3.2 Remarks on the construction of §3.1

Remark (1). If F is free, $0 \to \mathfrak{f} \to \mathbb{Z}F \to \mathbb{Z} \to 0$ is a free resolution of $\mathbb{Z}$ over $\mathbb{Z}F$ and so $H^k(F,A) = H_k(F,A) = 0$ for all A and all $k \geq 2$. This can be interpreted as case G = F of Theorem 2.

Remark (2). We obtain a "left" version of Theorem 2 by using $\mathfrak{r}^{n-1}\mathfrak{f} / \mathfrak{r}^n\mathfrak{f}$.

Remark (3). Theorem 2 remains true if $\mathbb{Z}$ is replaced by any given commutative ring K.

Remark (4). The argument in Theorem 1 shows, more generally, that if H is any subgroup, not necessarily normal, and $\mathfrak{h}$ is the right ideal generated by all 1-h, $h \in H$, then $\mathfrak{h}$ is free over $\mathbb{Z}F$ on 1-Y, where Y freely generates H.

Hence, if S is any subset of F, the right ideal generated by 1-S is free (on 1-Y, where Y is a free set of generators of $\langle S \rangle$).

Remark (5). (Kaplansky) If K is any ring, I, M are right ideals, both projective over K, and M is also a two sided-ideal, then IM is a projective right ideal: for

$$0 \to M \to K \to T \to 0$$

yields
$$0 \to I \otimes_K M \to I \to I \otimes_K T \to 0$$

(because I is projective), so that $IM \cong I \otimes_K M$. But $I \otimes_K M$ is projective.

Now assume I, M arise in the following way: Z is a cyclic K-module such that $\mathrm{Ker}(K \to Z) = I$ is projective; S is an image ring of K: $0 \to M \to K \to S \to 0$, where M is projective as right ideal and $M \leq I$. Then $K \to Z$ induces $S \to Z$ and

$$\ldots \to IM/IM^2 \to M/M^2 \to I/IM \to S \to Z \to 0$$

is a projective resolution of Z over S.

Remark (6). The following result is sometimes convenient.

PROPOSITION 2. If B is a subset of R such that $\{bR', b\in B\}$ is a basis of R/R', then $\mathfrak{f}/\mathfrak{f}^2$ is $\mathbb{Z}G$-free on all $(1-b) + \mathfrak{f}^2$, $b\in B$. (Note that B need not generate R.)

Proof. Generation. Since $R = \langle B, R' \rangle$, $\mathfrak{f}$ is the right ideal on all $1-b$, $b\in B$, and $xy-yx$, $x,y\in R$ (cf. Theorem 1, generation). As $xy-yx \in \mathfrak{f}^2$, $(1-B)+\mathfrak{f}^2$ generates $\mathfrak{f}/\mathfrak{f}^2$ as $\mathbb{Z}G$-module.

Freeness. Let $a = \Sigma (1-b_i)u_i \in \mathfrak{f}^2$, where $u_i \in \mathbb{Z}F$. If T is a transversal of R in F and

$$u_i = \Sigma v_{ij}t_j \quad , \quad v_{ij} \in \mathbb{Z}R,$$

then

$$\sum_{i,j} (1-b_i)v_{ij}t_j = \sum_{r,s} (1-y_r)(1-y_s)w_{rs} \quad , \quad w_{rs}\in \mathbb{Z}F,$$

where R is free on Y (by Theorem 1).

Equate coefficients of t_j and so for each j we have

$$\Sigma (1-b_i)v_{ij} \in \mathfrak{f}_0^2 \quad ,$$

where $\mathfrak{r}_o$ = augmentation ideal of $\mathbb{Z}R$. Hence, if η is the augmentation of $\mathbb{Z}R$,

$$\Sigma\ (1-b_i)(v_{ij}\eta)\ \epsilon\ \mathfrak{r}_o^2\ .$$

Now $\mathfrak{r}_o/\ \mathfrak{r}_o^2\ \cong R/R'$. Therefore

$$\prod_i b_i(v_{ij}\eta)\ \epsilon\ R'\ ,$$

whence $v_{ij}\eta = 0$ for all i (and all j). Thus $v_{ij}\ \epsilon\ \mathfrak{r}_o$ and therefore $u_i\ \epsilon\ \mathfrak{r}$, all i, as required.

Remark (7). The module $\mathfrak{r}/\ \mathfrak{r}^2$ is usually unnecessarily big. Suppose $R = \langle\ w^{-1}Zw,\ w\epsilon F\ \rangle$ (i.e., the normal closure of Z). Let $\mathfrak{a}$ be the right ideal generated by 1-Z. Then $\mathfrak{a}$ is a free right ideal (Remark (4) above) and we assert that the following is exact:

$$0 \to \mathfrak{a} \cap \mathfrak{f}\mathfrak{r}/\mathfrak{a}\mathfrak{r} \to \mathfrak{a}/\mathfrak{a}\mathfrak{r} \to \mathfrak{f}/\mathfrak{f}\mathfrak{r} \to \mathfrak{g} \to 0 \qquad (*)$$

with $\mathfrak{a}/\mathfrak{a}\mathfrak{r}$ $\mathbb{Z}G$-free.

All is clear except that $\mathfrak{a}/\mathfrak{a}\mathfrak{r} \to \mathfrak{r}/\mathfrak{f}\mathfrak{r}$ is surjective: i.e., that $\mathfrak{r}/\mathfrak{f}\mathfrak{r}$ is the G-module generated by the cosets of all 1-z, $z\epsilon Z$. Since R/R' is the G-module generated by all zR', our required conclusion is a consequence of the G-isomorphism

$$R/R' \xrightarrow{\sim} \mathfrak{r}/\mathfrak{f}\mathfrak{r}$$
$$rR' \to (1-r)+\mathfrak{f}\mathfrak{r}\ .$$

To establish this we need

LEMMA 5. If $\mathfrak{a}$ is a left ideal, free on a set S, then $\mathfrak{a}/\mathfrak{f}\mathfrak{a}$ is free as $\mathbb{Z}$-module on all $s+\mathfrak{f}\mathfrak{a}$, $s\epsilon S$. (Ditto with right if one uses $\mathfrak{a}\mathfrak{f}$.)

Proof. Take G = 1 in Lemma 4 (i) and switch from right to left.

Hence $\mathfrak{r}/\mathfrak{f}\mathfrak{r}$ is free abelian on $(1-y)+\mathfrak{f}\mathfrak{r}$, $y \in Y$ (using also the left version of Theorem 1). Thus $yR' \to (1-y)+\mathfrak{f}\mathfrak{r}$ yields an isomorphism $R/R' \to \mathfrak{r}/\mathfrak{f}\mathfrak{r}$ and clearly $rR' \to (1-r)+\mathfrak{f}\mathfrak{r}$ for all r in R. It is a G-isomorphism because

$$w^{-1}rwR' \to (1-w^{-1}rw) + \mathfrak{f}\mathfrak{r} = (1-r)w + \mathfrak{f}\mathfrak{r} .$$

We note that this G-isomorphism is a special case of

PROPOSITION 3. $\mathfrak{r}^n/\mathfrak{f}\mathfrak{r}^n \cong \underbrace{R/R' \otimes_{\mathbb{Z}} \dots \otimes_{\mathbb{Z}} R/R'}_{n}$, as G-modules, where G acts diagonally on the right.

Proof = exercise.

Observe that $\mathfrak{r}^n/\mathfrak{f}\mathfrak{r}^n$ is the kernel of $\mathfrak{f}\mathfrak{r}^{n-1}/\mathfrak{f}\mathfrak{r}^n \to \mathfrak{r}^n/\mathfrak{r}^{n+1}$ in our resolution.

Problem: How does one extend the sequence (*) above to a nice small free resolution? This is certainly possible when $|Z| = 1$: cf. groups with one defining relation in chapter 8.

Remark (8). The free resolution of Theorem 2 is finitely generated (i.e., each free module in it is finitely generated) if, and only if, F is finitely generated and G = F or G is finite.

For G finite and F finitely generated imply R finitely generated and so $\mathfrak{f}$, $\mathfrak{r}$ are both finitely generated.

Conversely, if $\mathfrak{f}/\mathfrak{f}\mathfrak{r}$ is finitely generated, F must be finitely generated; and if $\mathfrak{r}/\mathfrak{r}^2$ is also finitely generated, R is finitely generated. This implies G is finite, by a theorem of Schreier [6] (or cf. a slightly more general result by Karrass and Solitar [4]).

3.3 Cyclic groups.

Let G be cyclic of finite order h and $1 \to R \to F \to G \to 1$ the simplest presentation : i.e., F is free on x, $R = \langle x^h \rangle$. Suppose $x \to \bar{x}$ in G.

If the corresponding resolution, according to Theorem 2, is

$$\cdots \to X_2 \to X_1 \to X_0 \to \mathbb{Z} \to 0,$$

then X_{2n} is free on $x_{2n} = (1-x^h)^n + \mathfrak{r}^{n+1}$;

X_{2n-1} is free on $x_{2n-1} = (1-x)(1-x^h)^{n-1} + \mathfrak{f}\mathfrak{r}^n$.

As $$(1-x^h)^n = (1-x)(1-x^h)^{n-1}(1 + x + \cdots + x^{h-1}),$$

$$x_{2n+1} \to x_{2n}(1-\bar{x})$$

$$x_{2n} \to x_{2n-1}\tau ,$$

where $\tau = 1+\bar{x}+\cdots+\bar{x}^{h-1}$.

Now $x_1 \to 1$ yields $X_1 \xrightarrow{\sim} \mathbb{Z}G$ and then $X_{2n+1} \to X_{2n} \to X_{2n-1}$ becomes $\mathbb{Z}G \xrightarrow{1-\bar{x}} \mathbb{Z}G \xrightarrow{\tau} \mathbb{Z}G$. So our resolution becomes

$$\cdots \to \mathbb{Z}G \xrightarrow{\tau} \mathbb{Z}G \xrightarrow{1-\bar{x}} \mathbb{Z}G \xrightarrow{\tau} \mathbb{Z}G \xrightarrow{1-\bar{x}} \mathbb{Z}G \to \mathbb{Z} \to 0.$$

For any right module A,

$$0 \to A^G \to A \xrightarrow{1-\bar{x}} A \xrightarrow{\tau} A \xrightarrow{1-\bar{x}} A \to \cdots$$

and so, for all $n \geq 0$,

$$\underline{H^{2n+1}(G,A) = A_\tau/A\mathfrak{g}}$$

$$\underline{H^{2n+2}(G,A) = A^G/A\tau}$$

where $A_\tau = \operatorname{Ker} \tau$.

Further,

$$\underline{H_k(G,A) = H^{k+1}(G,A)}, \quad \text{all } k \geq 1.$$

If A is trivial (i.e., $A^G = A$), then the above formulae become

$$H^{2n+1}(G,A) = A_h$$

$$H^{2n+2}(G,A) = A/hA.$$

Definition. If both $A_\tau/A\mathfrak{g}$ and $A^G/A\tau$ are finite groups we call

$$h(A) = \frac{(A^G:A\tau)}{(A_\tau:A\mathfrak{g})} = \frac{|H^2(G,A)|}{|H^1(G,A)|}$$

the Herbrand quotient of A.

LEMMA 6. If A is finite, h(A) = 1.

Proof. $0 \to A^G \to A \xrightarrow{1-\bar{x}} A \to A_G \to 0$ is exact and so $|A^G| = |A_G|$.

Then

$$0 \to A_\tau/A\mathfrak{g} \to A_G \to A^G \to A^G/A\tau \to 0$$

is exact, whence the result.

3.4 The standard resolution.

Let G be given and take F to be free on $\{x_g;\ g\in G^*\}$, where G^* is the set of all non-identity elements of G. Suppose π is the homomorphism extending $x_g \to g$, $g\in G^*$. The presentation

$$1 \to R \to F \overset{\pi}{\to} G \to 1$$

with this F and this π we shall call the <u>standard presentation</u> and the corresponding resolution we call the <u>standard resolution</u> for G. We show that this is, in fact, the "standard" or "bar" resolution to be found in the literature.

Let us write x_1 for 1 in F. Then $\{x_g;\ g\in G\}$ is a Schreier transversal for R in F and so R is freely generated by all $y_{a,b} = x_a x_b x_{ab}^{-1}$, $a,b\in G^*$. (Cf. Rotman, p.242, for a good account of this theory, based on work of A.J. Weir.) Hence $\{1 - y_{a,b};\ a,b\in G^*\}$ is a set of free generators of $\mathfrak{r}$, as right F-module. We set

$$(a,b) = x_{ab} - x_a x_b = (1 - y_{a,b})x_{ab}$$

for all a, b in G. Clearly $(a,b) \neq 0$ if, and only if, $a,b\in G^*$ and $\{(a,b);\ a,b\in G^*\}$ is also a set of free generators of $\mathfrak{r}$.

Now define

$$((g_1,\dots,g_{2n})) = (g_1,g_2)(g_3,g_4)\dots(g_{2n-1},g_{2n}) + \mathfrak{r}^{n+1},$$

$$((g_1,\dots,g_{2n-1})) = (1 - x_{g_1})(g_2,g_3)\dots(g_{2n-2},\ g_{2n-1}) + \mathfrak{f}\mathfrak{r}^n$$

for all $n \geq 1$ and all $g_1,\dots,g_{2n}$ in G. If

$$X_{2n} = \mathfrak{r}^n/\mathfrak{r}^{n+1}, \quad X_{2n-1} = \mathfrak{f}\mathfrak{r}^{n-1}/\mathfrak{f}\mathfrak{r}^n \quad (n \geq 1),$$

then, for all $k \geq 1$, X_k is freely generated as right G-module by all non-zero elements $((g_1,\dots,g_k))$, for g_i in G, $i = 1,\dots,k$. (Of course, $((g_1,\dots,g_k)) = 0$ if, and only if, at least one g_i is 1.)

PROPOSITION 4. If d_k is the mapping $X_k \to X_{k-1}$ in the standard resolution for G, then for all $k > 1$,

$$((g_1,\dots,g_k))\, d_k = ((g_2,\dots,g_k)) + \sum_{j=1}^{k-1} (-1)^j ((g_1,\dots,g_j g_{j+1},\dots,g_k))$$

$$+ (-1)^k\, ((g_1,\dots,g_{k-1}))\, g_k.$$

We remark that if we let $X_0 = \mathbb{Z}G$ and interpret $((g_1,\dots,g_k))$ with $k = 0$ to be the identity element of $\mathbb{Z}G$, then Proposition 4 remains true for $k = 1$.

The case $k = 2$ is a consequence of the identity

$$(a,b) = (1-x_b) - (1 - x_{ab}) + (1 - x_a)x_b.$$

To see the formula with $k = 3$, we first obtain

$$x_a y_{b,c} = y_{a,b}\, y_{ab,c}\, y_{a,bc}^{-1}\, x_a$$

by calculating $x_a(x_b x_c)$, $(x_a x_b)x_c$ and equating. Then

$$(1 - x_a)(b,c) = (b,c) - x_a(1 - y_{b,c})x_{bc}$$

$$= (b,c) - (1 - y_{a,b}y_{ab,c}y_{a,bc}^{-1})x_a x_{bc}, \text{ by the above,}$$

$$\equiv (b,c) - (a,b)x_c - (ab,c) + (a,bc), \bmod \mathfrak{X}^2 ,$$

thus giving Proposition 4 with k = 3.

We complete the proof by induction on k. Write $(g) = 1-x_g$. Suppose we have the result for all $\ell < k$ and consider $\ell = k$.

<u>(1) k even</u>: k = 2n.

$$(g_1,g_2)\ldots(g_{2n-1},g_{2n}) = \{(g_2) - (g_1g_2) - (g_1)(g_2) + (g_1)\}(g_3,g_4)\ldots$$

by k = 2. We replace the (g_1)-multiple $(g_2)(g_3,g_4)\ldots(g_{2n-1},g_{2n})$ in the third term by the appropriate element in $\mathfrak{X}^{n-1}$ (case k = 2n-1) and then our formula is established.

<u>(2) k odd</u>: k = 2n+1. By k = 3, we have, modulo $\mathfrak{X}^{n+1}$,

$$(g_1)(g_2,g_3)\ldots(g_{2n},g_{2n+1})$$

$$\equiv \underbrace{\{(g_2,g_3)-(g_1g_2,g_3)+(g_1,g_2g_3)}_{\text{correct first three terms}}-(g_1,g_2)x_{g_3}\}(g_4,g_5)\ldots$$

and

$$-(g_1,g_2)x_{g_3}(g_4,g_5)\ldots = \{(g_1,g_2)(g_3)-(g_1,g_2)\}(g_4,g_5)\ldots \quad .$$

In the right hand side here we replace $(g_3)(g_4,g_5)\ldots$ according to the formula for k = 2n-1 and complete the induction.

<u>Terminology</u>. Cochains, cocycles, coboundaries arising from a standard resolution have the adjective "standard" prefixed to them.

Standard 2-cocycles are often called <u>factor-sets</u>.

3.5 $H^1(G, \ )$ and $H_1(G, \ )$.

By Proposition 2 of §2.2, for any left G-module A, we have the exact sequence

$$0 \to H_1(G,A) \to \mathfrak{g}\otimes_G A \to \mathbb{Z}G\otimes_G A.$$

Hence

$$H_1(G,A) = \mathrm{Ker}(\mathfrak{g}\otimes_G A \to A).$$

Assume that G acts trivially. Then $H_1(G,A) = \mathfrak{g}\otimes_G A$. Now $\mathfrak{g}\otimes_G A = (\mathfrak{g}\otimes_G \mathbb{Z})\otimes_{\mathbb{Z}} A$ and $\mathfrak{g}\otimes_G \mathbb{Z} \cong \mathfrak{g}/\mathfrak{g}^2$ (cf. Theorem 2 in chapter 2) and $\mathfrak{g}/\mathfrak{g}^2 \cong G/G'$. (§2.4). So we have,

<u>if A is trivial, $H_1(G,A) \cong G/G'\otimes_{\mathbb{Z}} A$.</u>

Proposition 1 of §2.1 yields the exact sequence

$$\mathrm{Hom}_G(\mathbb{Z}G, A) \to \mathrm{Hom}_G(\mathfrak{g}, A) \to H^1(G,A) \to 0.$$

The one-one correspondence of Lemma 1 above between $\mathrm{Hom}_G(\mathfrak{g}, A)$ and the set Der(G,A) of all derivations of G in A is easily seen to be a group isomorphism. In this isomorphism, what is the subgroup corresponding to the image of $\mathrm{Hom}_G(\mathbb{Z}G,A)$? A homomorphism $\mathfrak{g} \to A$ comes from one of $\mathbb{Z}G$ in A if, and only if, it is of the form

$$1-g \to a(1-g)$$

for some fixed a in A. We call the corresponding derivation: $g \to a(1-g)$ an <u>inner derivation</u> (or <u>principal crossed homomorphism)</u> and write the group of these $\mathrm{Ider}(G,A)$. Clearly

$$\mathrm{Ider}(G,A) \cong A/A^G.$$

So we have

$$\underline{H^1(G,A) \cong \mathrm{Der}(G,A)/\mathrm{Ider}(G,A)}$$

and the exact sequence

$$0 \to A^G \to A \to \mathrm{Der}(G,A) \to H^1(G,A) \to 0.$$

There is a very useful group theoretic interpretation of $H^1(G,A)$. Consider a group E and a normal abelian subgroup A of E. Let D be the subgroup of Aut E consisting of all α fixing A and E/A elementwise: i.e., D consists of all α such that

$$[A,\alpha] = 1 \quad \text{and} \quad [E,\alpha] \leq A.$$

Clearly, for each α in D, α': $e \to [e,\alpha]$ is a derivation of E in A. Now

$$[ea,\alpha] = [e,\alpha] \quad \text{for all a in A,}$$

and hence α' yields a derivation α'' of $G = E/A$ in A. It is trivial to verify that $\alpha \to \alpha''$ is an isomorphism of abelian groups:

$$D \xrightarrow{\sim} \mathrm{Der}(G,A).$$

Let D_0 be the subgroup of D corresponding to $\mathrm{Ider}(G,A)$. So $\alpha \in D_0$ if, and only if, there exists a in A so that, for all e,

$$[e,\alpha] = a\,(a^{-1})^e = [e,a],$$

i.e., $$e^\alpha = e^a,$$

i.e., α is conjugation by a. Under the natural homomorphism $E \to \mathrm{Aut}\, E$, $A \to D$ and the image is precisely D_0. Thus we have

<u>PROPOSITION 5. If $A \lhd E$ and A is abelian, then $H^1(E/A,A)$ is isomorphic to the group of all automorphisms of E leaving A and E/A elementwise fixed modulo the image of A in Aut E.</u>

3.6 $H^2(G,\ )$ and $H_2(G,\ )$.

Again Proposition 1 of §2.1 gives (when applied to the resolution of Theorem 2) the exact sequence

$$\mathrm{Hom}_G(\mathfrak{f}/\mathfrak{f}\mathfrak{r},A) \xrightarrow{\mathrm{Res}} \mathrm{Hom}_G(\mathfrak{r}/\mathfrak{f}\mathfrak{r},A) \to H^2(G,A) \to 0, \qquad (*)$$

where Res = restriction.

By Proposition 3, $\mathrm{Hom}_G(\mathfrak{r}/\mathfrak{f}\mathfrak{r},A) \cong \mathrm{Hom}_G(R/R',A)$. If we view A as F-module via $F \to G$, then

$$\begin{aligned}\mathrm{Hom}_G(\mathfrak{f}/\mathfrak{f}\mathfrak{r},A) &\cong \mathrm{Hom}_F(\mathfrak{f},A) && \text{(additive groups)}\\ &\cong \mathrm{Der}\,(F,A) && \text{(Lemma 1).}\end{aligned}$$

So (*) yields

PROPOSITION 6. (Mac Lane [5].) For any G-module A,

$$\mathrm{Der}\,(F,A) \xrightarrow{\mathrm{Res}} \mathrm{Hom}_G(R/R',A) \to H^2(G,A) \to 0$$

is exact.

Now consider Proposition 2 of chapter 2 (§2.2) with q = 2 and $A = \mathbb{Z}$:

$$H_2(G,\mathbb{Z}) = \mathrm{Ker}\,(\mathfrak{r}/\mathfrak{f}\mathfrak{r} \otimes_G \mathbb{Z} \to \mathfrak{f}/\mathfrak{f}\mathfrak{r} \otimes_G \mathbb{Z}).$$

But

$$\mathfrak{r}/\mathfrak{f}\mathfrak{r} \otimes_G \mathbb{Z} \cong R/R' \otimes_G \mathbb{Z} \cong (R/R')/(R/R')\mathfrak{g} \cong R/[R,F];$$

and

$$\mathfrak{f}/\mathfrak{f}\mathfrak{r} \otimes_G \mathbb{Z} \cong (\mathfrak{f}/\mathfrak{f}\mathfrak{r})/(\mathfrak{f}/\mathfrak{f}\mathfrak{r})\mathfrak{g} \cong \mathfrak{f}/\mathfrak{f}^2 \cong F/F'.$$

So

$$H_2(G,\mathbb{Z}) = \mathrm{Ker}\,(R/[R,F] \to F/F') = R \cap F'/[R,F].$$

PROPOSITION 7. (Hopf [3].) $H_2(G,\mathbb{Z}) \cong R \cap F'/[R,F]$.

A special case of Propositions 6 and 7 appeared in a veiled form in the important work of Schur on projective representations of finite groups [7]. The study of these is thrown back to ordinary representation theory via the central presentations of the group and it was these latter objects that naturally led Schur to the group $H^2(G,\mathbb{C}^*)$, with G acting trivially on $\mathbb{C}^*$. This group he called the multiplicator of G. (Cf. also §9.9).

Exercise. A finitely generated abelian group G is cyclic if, and only if, $H_2(G,\mathbb{Z}) = 0$.

3.7 The universal coefficient theorem for cohomology.

The argument leading to Proposition 7 also yields

$$H_2(G,\mathbb{Z}) = \frac{\mathfrak{r} \cap \mathfrak{f}^2}{\mathfrak{f}\mathfrak{r} + \mathfrak{r}\mathfrak{f}} .$$

This formula can be generalized. (Cf. [2] for what follows.)

By Proposition 2 (§2.2), if $n \geq 1$,

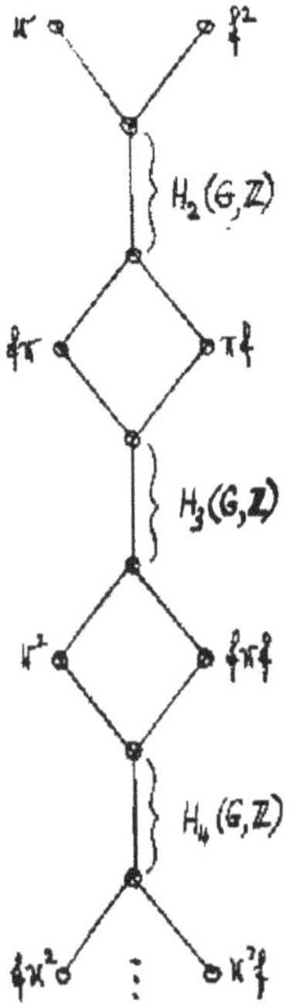

$$H_{2n}(G,\mathbb{Z}) = \mathrm{Ker}((\mathfrak{r}^n/\mathfrak{f}\mathfrak{r}^n)_G \to (\mathfrak{f}\mathfrak{r}^{n-1}/\mathfrak{f}\mathfrak{r}^n)_G)$$
$$= \mathrm{Ker}(\mathfrak{r}^n/\mathfrak{f}\mathfrak{r}^n + \mathfrak{r}^n\mathfrak{f} \to \mathfrak{f}\mathfrak{r}^{n-1}/\mathfrak{f}\mathfrak{r}^{n-1}\mathfrak{f})$$

and so

$$\underline{H_{2n}(G,\mathbb{Z}) = \mathfrak{r}^n \cap \mathfrak{f}\mathfrak{r}^{n-1}\mathfrak{f}/\mathfrak{f}\mathfrak{r}^n + \mathfrak{r}^n\mathfrak{f}} ;$$

while

$$H_{2n+1}(G,\mathbb{Z}) = \mathrm{Ker}((\mathfrak{f}\mathfrak{r}^n/\mathfrak{r}^{n+1})_G \to (\mathfrak{r}^n/\mathfrak{r}^{n+1})_G)$$
$$= \mathrm{Ker}(\mathfrak{f}\mathfrak{r}^n/\mathfrak{f}\mathfrak{r}^n\mathfrak{f} + \mathfrak{r}^{n+1} \to \mathfrak{r}^n/\mathfrak{r}^n\mathfrak{f}),$$

so that

$$\underline{H_{2n+1}(G,\mathbb{Z}) = \mathfrak{f}\mathfrak{r}^n \cap \mathfrak{r}^n\mathfrak{f}/\mathfrak{r}^{n+1} + \mathfrak{f}\mathfrak{r}^n\mathfrak{f}} .$$

To remember these formulae, label the left hand vertices of the above diagram with the numerators of the resolution; then do the same for the right hand vertices but transpose down one and multiply on the right by $\mathfrak{f}$.

As an application of these formulae we now prove

<u>THEOREM 3. (The Universal Coefficient Theorem for Cohomology.)</u>
<u>For all $q \geq 1$ and every trivial G-module T,</u>

$$H^q(G,T) \cong \mathrm{Hom}_{\mathbb{Z}}(H_q(G,\mathbb{Z}),\ T) \oplus \mathrm{Ext}^1_{\mathbb{Z}}(\ H_{q-1}(G,\mathbb{Z}),\ T).$$

<u>Proof</u>. By Proposition 1 of §2.1 (applied to the resolution of Theorem 2)

$$H^{2n}(G,T) \cong \mathrm{Coker}(\ \mathrm{Hom}_G(\mathfrak{f}\mathfrak{r}^{n-1}/\mathfrak{f}\mathfrak{r}^n,\ T) \to \mathrm{Hom}_G(\mathfrak{r}^n/\mathfrak{f}\mathfrak{r}^n,\ T))$$

$$= \mathrm{Coker}(\ \mathrm{Hom}(\ \mathfrak{f}\mathfrak{r}^{n-1}/\mathfrak{f}\mathfrak{r}^{n-1}\mathfrak{f}\ ,T) \to \mathrm{Hom}(\ \mathfrak{r}^n/\mathfrak{f}\mathfrak{r}^n+\mathfrak{r}^n\mathfrak{f},T)),$$

since G acts trivially. (Hom = $\mathrm{Hom}_{\mathbb{Z}}$.) Now

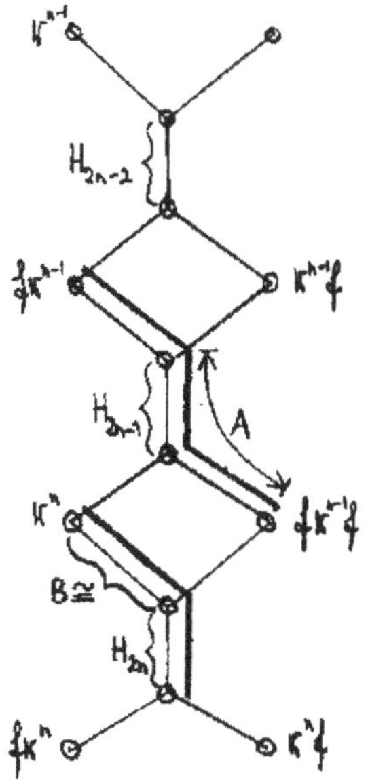

$\mathfrak{f}\mathfrak{r}^{n-1}/\mathfrak{f}\mathfrak{r}^{n-1} \cap \mathfrak{r}^{n-1}\mathfrak{f} \cong$ a subgroup of $\mathfrak{r}^{n-1}/\mathfrak{r}^{n-1}\mathfrak{f}$ and this is $\mathbb{Z}$-free (Lemma 5). So $\mathfrak{f}\mathfrak{r}^{n-1}/\mathfrak{f}\mathfrak{r}^{n-1}\mathfrak{f}$ splits over A (= group shown in the diagram), i.e., is isomorphic to A ⊕ C, say. Again, $\mathfrak{r}^n/\mathfrak{r}^n \cap \mathfrak{f}\mathfrak{r}^{n-1}\mathfrak{f}$ is free (because it is isomorphic to a subgroup of the free group A) and so it also splits and is isomorphic to $H_{2n} \oplus B$.

Consequently

$H^{2n}(G,T)$

$$\cong \mathrm{Hom}(H_{2n},T) \oplus \mathrm{Coker}(\mathrm{Hom}(A,T) \to \mathrm{Hom}(B,T)).$$

Now $0 \to B \to A \to H_{2n-1} \to 0$ is exact and as A is $\mathbb{Z}$-free, $\mathrm{Ext}^1_{\mathbb{Z}}(A,T) = 0$. So

$$\mathrm{Coker}(\mathrm{Hom}(A,T) \to \mathrm{Hom}\ (B,T))$$
$$\cong \mathrm{Ext}^1_{\mathbb{Z}}(\ H_{2n-1}(G,\mathbb{Z}),\ T).$$

The proof for odd $q \geq 1$ is similar; for q=1 the result is clear.

References

[1] Gruenberg, K.W.: Resolutions by relations, J.London Math. Soc., 35 (1960) 481-494.

[2] Gruenberg, K.W.: The universal coefficient theorem in the cohomology of groups, J. London Math. Soc., 43 (1968).

[3] Hopf, H.: Fundamentalgruppe und zweite Bettische Gruppe, Comm. Math. Helvetici 14 (1941/42) 257-309.

[4] Karrass, A. and Solitar, D.: Note on a theorem of Schreier, Proc. Amer. Math. Soc. 8 (1957) 696-697.

[5] Mac Lane, S.: Cohomology theory in abstract groups, III, Annals Math. 50 (1949) 736-761.

[6] Schreier, O.: Die Untergruppen der freien Gruppen, Abh. Math. Sem. Hamburg 5 (1927) 161-183.

[7] Schur, I.: Über die Darstellungen der endlichen Gruppen durch gebrochene lineare Substitutionen, Crelle 127 (1904) 20-50.

CHAPTER 4

FREE GROUPS

4.1 Dimension subgroups.

We saw that $H_2(G,\mathbb{Z})$ has a group theoretic expression as well as a ring theoretic one. Can we translate the formulae for $H_k(G,\mathbb{Z})$, $k \geq 2$, given in §3.7, into group theoretic form? The answer is yes; but it does not seem possible to do this within F itself. We wish to indicate why this is likely to be so (Proposition 1, below).

Suppose G is an arbitrary group and $\mathfrak{m}$ a right ideal in $\mathbb{Z}G$. Then $M = (1+\mathfrak{m}) \cap G$ is a subgroup of G. If $\mathfrak{m}$ is two sided, M is normal. The problem of identifying M is usually difficult, even if $\mathfrak{m}$ is an "easy" ideal. Of course, if $\mathfrak{m}$ is the right ideal generated by a set 1-S, where $S \leq G$, then clearly $M = \langle S \rangle$.

<u>Definition</u>. $D_n(G) = (1+\mathfrak{g}^n) \cap G$ is called the n-th <u>dimension group</u> of G.

Clearly, $D_1(G) = G$ and $D_2(G) = G_2$ (because

$$(1-w)+\mathfrak{g}^2 \to wG'$$

is an isomorphism of $\mathfrak{g}/\mathfrak{g}^2$ onto G/G': cf. §2.4). It is also clear that $D_n(G) \geq G_n$ (use induction on n).

Open problem: Is $D_n(G) = G_n$ for all n?.

This is an exceedingly subtle problem and relatively little progress has been made. It was shown recently that $D_3(G) = G_3$ (due to G. Higman and D. Rees, independently); and $D_4(G) = G_4$ provided no subquotient of G is a finite 2-group of class 3 (I.B.S. Passi). Cf. [13],[14].

It is easy to see that the problem need only be solved for finite p-groups. We argue thus. If the result holds for all finitely generated groups then it holds for all groups. Further, if $K \triangleleft G$, $K \leq G_n$ and $D_n(G/K) = G_n/K$, then $D_n(G) = G_n$. Hence one only needs to consider finitely generated nilpotent groups. Such groups are known to possess a family of normal subgroups intersecting in 1 and whose quotient groups are finite prime power groups. Hence our assertion.

Dimension subgroups can also be defined when the coefficient ring is an arbitrary commutative ring K. One might then denote the groups by $D_n(KG)$ (so that $D_n(G) = D_n(\mathbb{Z}G)$). Changing K can drastically change the corresponding dimension groups. We cite one result (Jennings [5]; cf. also P. Hall's Canadian notes, chapter 7): if K is a field of characteristic zero, $D_n(KG)$ is the torsion group of G modulo G_n (i.e., $\{x \in G \mid xG_n$ has finite order$\}$).

Dimension groups over fields of finite characteristic have been studied by many authors. Cf., in particular, Zassenhaus [12], Jennings [4], Lazard [6].

In the integral case the only general result is the following.

<u>MAGNUS' THEOREM</u> [8]. <u>If F is free, $D_n(F) = F_n$ for all $n \geq 1$.</u>

Using this, we can identify the ideals appearing in the resolution of the last chapter.

<u>PROPOSITION 1. If F is free, R is normal in F and $\mathfrak{r}$ is the right ideal on $1-R$, then for all $n \geq 1$,</u>

$$(\mathfrak{f}\mathfrak{r}^n + 1) \cap F = (\mathfrak{r}^{n+1} + 1) \cap F = R_{n+1}.$$

(The case $n = 1$ is an old result of <u>Schumann</u> [11]. Cf. also Fox [2].)

<u>Proof</u>. Observe first that if $\mathfrak{a}_0$ is a right ideal of $\mathbb{Z}R$ and $\mathfrak{a}_0$ generates the right ideal $\mathfrak{a}$ of $\mathbb{Z}F$, then $\mathfrak{a} \cap \mathbb{Z}R = \mathfrak{a}_0$.

Hence, if $\mathfrak{r}_0$ is the augmentation ideal of $\mathbb{Z}R$, we have $\mathfrak{r}^n \cap \mathbb{Z}R = \mathfrak{r}_0^n$. Now $w \in (\mathfrak{r}^n + 1) \cap F$ if, and only if, $1-w \in \mathfrak{r}^n$. As $w \in R$ (case $n = 1$), $1-w \in \mathfrak{r}^n \cap \mathbb{Z}R = \mathfrak{r}_0^n$ and so $w \in R_n$ by Magnus' Theorem.

Suppose $w \in (\mathfrak{f}\mathfrak{r}^n + 1) \cap F$. Now $w \in R$ and $1-w \in \mathfrak{r}_0^n$. So

$$1-w = \Sigma\, (1-y_{i_1}) \ldots (1-y_{i_n})\alpha_{\underline{i}}$$

with $\alpha_{\underline{i}} \in \mathbb{Z}R$. By Lemma 5 (§3.2), $\mathfrak{r}^n/\mathfrak{f}\mathfrak{r}^n$ is free abelian on all $(1-y_{i_1}) \ldots (1-y_{i_n})$ and so $1-w \in \mathfrak{f}\mathfrak{r}^n$ if, and only if, all $\alpha_{\underline{i}} \in \mathfrak{r}_0$: i.e., if, and only if, $1-w \in \mathfrak{r}_0^{n+1}$, i.e., if, and only if, $w \in R_{n+1}$.

4.2 Residual nilpotence of the augmentation ideal.

Let F be a free group on $(x_i)_{i\in I}$. Proposition 1 of §3.1 is a powerful tool for studing $\mathbb{Z}F$. The systematic use of derivations in this context is due to R.H. Fox [2].

Let ϵ be the augmentation $\mathbb{Z}F \to \mathbb{Z}$; and let d_i be the derivation $x_j \to \delta_{ij}$. Throughout this section we shall write $\xi_i = x_i - 1$.

For any α in $\mathbb{Z}F$, $\alpha - \alpha\epsilon$ is uniquely $\Sigma\, \xi_i \alpha_i$, where $\alpha_i \in \mathbb{Z}F$. Now α_i must be αd_i - cf. the connexion between derivations in $\mathbb{Z}F$ and $\mathrm{Hom}_F(\mathfrak{f}, \mathbb{Z}F)$ (§3.1)(the homomorphism corresponding to a derivation d takes $\alpha - \alpha\epsilon$ to $-\alpha d$). So

$$\alpha = \alpha\epsilon + \Sigma\, \xi_i(\alpha d_i).$$

Repeat:
$$\alpha = \alpha\epsilon + \Sigma\, \xi_i(\alpha d_i \epsilon) + \Sigma\, \xi_i \xi_j(\alpha d_i d_j);$$

and generally:

$$\alpha = \alpha\epsilon + \Sigma\, \xi_i(\alpha d_i \epsilon) + \ldots + \Sigma\, \xi_{i_1}\cdots\xi_{i_n}(\alpha d_{i_1}\cdots d_{i_n}\epsilon) + \alpha_{(n)} \quad (*)$$

where
$$\alpha_{(n)} = \Sigma\, \xi_{i_1}\cdots\xi_{i_{n+1}}\left(\alpha\, d_{i_1}\cdots d_{i_{n+1}}\right).$$

LEMMA 1. *$\alpha \in \mathfrak{f}^n$ if, and only if, $\alpha\, d_{i_1}\cdots d_{i_r}\epsilon = 0$, all $i_1,\ldots,i_r$ and all $r = 0, 1, \ldots, n-1$.*

Proof. Immediate from (*) above, Proposition 1 and Lemma 4 of §3.1.

<u>Definition</u>. If $\alpha = m_1w_1 + \ldots + m_kw_k$, $m_i \neq 0$ in $\mathbb{Z}$, $w_i \in F$ and $w_i \neq w_j$ if $i \neq j$, define the <u>length</u> of α to be $\ell(\alpha) = \max_i\{\ell(w_i)\}$. Put $\ell(0) = 0$.

<u>LEMMA 2. If $\alpha \neq 0$ and $\alpha \in \mathfrak{f}^n$ then $\ell(\alpha) \geq \frac{n}{2}$.</u>

Remark: n will not do instead of $\frac{n}{2}$: e.g. $(x-1)^2x^{-1} \in \mathfrak{f}^2$ but $\ell = 1$.

<u>Proof</u>. (Fox [2].) True if n = 1,2. Assume $\alpha \in \mathfrak{f}^n$ and $\ell(\alpha) = \ell < \frac{n}{2}$. Each element of F involved in α begins in some $x_i^{\pm 1}$: so

$$\alpha = \sum_i (x_i\beta_i + x_i^{-1}\gamma_i)$$

and $\ell(\beta_i)$, $\ell(\gamma_i) < \ell$. Then

$$\alpha d_i = \alpha_i - x_i^{-1}\gamma_i,$$

where $\ell(\alpha_i) < \ell$. If $j \neq i$, $\alpha d_i d_j = \alpha_i d_j - \gamma_i d_j$ has length $< \ell$;

and $$(x_i(\alpha d_i))d_i = \alpha_i + \alpha_i d_i - \gamma_i d_i$$
$$= \alpha d_i d_i + \alpha d_i \quad \text{has length} < \ell \text{ also.}$$

By Lemma 1, $\alpha \in \mathfrak{f}^n$ implies $\alpha d_i d_j \in \mathfrak{f}^{n-2}$ and $\alpha d_i \in \mathfrak{f}^{n-1}$. So, by induction hypothesis,

$$\alpha d_i d_j = 0 \quad \text{if} \quad i \neq j \; ; \tag{1}$$

and since $(x_i(\alpha d_i))d_i \in \mathfrak{f}^{n-2}$, induction again gives

$$\alpha d_i d_i + \alpha d_i = 0 \; . \tag{2}$$

(For by assumption, $\ell - 1 < \frac{n-2}{2}$ and by induction, the lengths of $\alpha d_i d_j$ (when $i \neq j$) and $\alpha d_i d_i + \alpha d_i$ should be $\geqslant \frac{n-2}{2}$ if they are non-zero.) Thus

$$\alpha d_i = \alpha d_i \varepsilon + \Sigma \xi_j(\alpha d_i d_j)$$
$$= (x_i - 1)\alpha d_i d_i \quad \text{by (1) and because } \alpha d_i \varepsilon \, \mathfrak{f},$$
$$= (1 - x_i)(\alpha d_i) \quad \text{by (2).}$$

Therefore $\alpha d_i = 0$. This holds for all i and therefore $\alpha = 0$.

THEOREM 1. (Magnus [7].) $\bigcap \mathfrak{f}^n = 0$.

Equivalently, if α β are elements of $\mathbf{Z}F$ so that $\alpha\varepsilon = \beta\varepsilon$, $\alpha d_i \varepsilon = \beta d_i \varepsilon$, $\alpha d_i d_j \varepsilon = \beta d_i d_j \varepsilon$, etc., then $\alpha = \beta$.

All this remains true when $\mathbf{Z}$ is replaced by an arbitrary coefficient ring.

Exercise. Let $1 \to R \to F \xrightarrow{\pi} G \to 1$ be a finite presentation of a group G: i.e., F is a free group on a finite set, say $\{x_1, \ldots, x_d\}$, and R is the normal closure of a finite set, say $\{z_1, \ldots, z_r\}$.

Prove that the r x d matrix $(z_i d_j \pi)$ over $\mathbf{Z}G$ corresponds to the $\mathbf{Z}G$-module $\mathfrak{g}$. If $\mathfrak{g}_2 = \text{Ker}\ \alpha$, where $\alpha: \mathbf{Z}G \to \mathbf{Z}(G/G') = K$, then $(z_i d_j \pi\alpha)$ is an r x d matrix over the commutative ring K and the corresponding K-module is $\mathfrak{g}/\mathfrak{g}_2$. Thus $(z_i d_j \pi\alpha)$ is an invariant of G. It is called the Alexander matrix of the presentation. (Cf. [1], chapter 7, for another proof.) (Hint: cf. Remark 7, §3.2.)

This theory has also been lucidly discussed in a recent paper by J. Gamst, Linearisierung von Gruppendaten mit Anwendungen auf Knotengruppen, Math.Zeit. 97 (1967) 291-302.

4.3 Residual properties of free groups.

Definition. Given any class of groups $\mathfrak{X}$ (closed under isomorphisms and containing 1), we define a new class, called residually-$\mathfrak{X}$, written $R\mathfrak{X}$ as follows: $G \in R\mathfrak{X}$ if, and only if, to each $g \neq 1$ in G, there corresponds a normal N so that $g \notin N$ and $G/N \in \mathfrak{X}$. Equivalently, $G \in R\mathfrak{X}$ if, and only if, there exists a family (N_i) of normal subgroups of G so that $\bigcap N_i = 1$ and $G/N_i \in \mathfrak{X}$ for all i. Note that $\mathfrak{X} \leq R\mathfrak{X} = R^2\mathfrak{X}$.

Theorem 1 above implies $\bigcap D_n(F) = 1$. Since $D_n(F) \geq F_n$ for all n (equality actually), this shows that F is residually nilpotent.

More is true. Theorem 1 with $\mathbb{Z}$ replaced by $\mathbb{F}_p$ shows that $\bigcap D_n(\mathbb{F}_p G) = 1$. Now $(1-w)^p = 1-w^p$ implies $D_n{}^p \leq D_{n+1}$ for all $n \geq 1$ $(D_i = D_i(\mathbb{F}_p G))$ and so F/D_n has finite p-exponent. Hence, if F is finitely generated, each F/D_n is a finite p-group. Since every free group is residually finitely generated, we have

THEOREM 2. Free groups are residually finite p-groups, for all primes p.

It is worth noting that the argument leading to Theorem 2 is quite general. Let us temporarily write the augmentation ideal of KG as $\mathfrak{g}_K$. Thus $\mathfrak{g}_{\mathbb{Z}} = \mathfrak{g}$.

PROPOSITION 2. *(i) If $\bigcap \mathfrak{g}_{\mathbb{Z}}^{\,n} = 0$, then G is residually nilpotent.*

(ii) If $\bigcap \mathfrak{g}_{\mathbb{F}_p}^{\,n} = 0$, then G is residually "nilpotent and of finite p-exponent".

The converse of (i) is false. Any abelian divisible torsion group is a counterexample in view of the simple

PROPOSITION 3. *If G is any group so that G/G' is divisible and torsion, then $\mathfrak{g}^2 = \mathfrak{g}^3$.*

Proof. For any x in G and any positive integer n,

$$n(x-1) \equiv x^n - 1 \mod \mathfrak{g}^2.$$

So, if $x^k \in G'$ and y is any element in G,

$$k(x-1)(y-1) \equiv (x-1)(y^k-1) \equiv 0 \mod \mathfrak{g}^3.$$

Take any z in G and find y so that $y^k \equiv z \mod G'$. Then

$$y^k - 1 \equiv z-1 \mod \mathfrak{g}^2$$

and hence

$$(x-1)(z-1) \equiv \mod \mathfrak{g}^3.$$

The converse of (ii) is true. This is a theorem of Mal'cev [10]. A non-trivial generalization of this theorem is the following result.

THEOREM. *Let K be any commutative ring with a maximal ideal $\mathfrak{m}$ so that $\bigcap \mathfrak{m}^n = 0$, $p \in \mathfrak{m}$ and $p \notin \mathfrak{m}^2$ if $\mathfrak{m} \neq 0$. If G is residually "nilpotent and of finite p-exponent", then* $\bigcap \mathfrak{g}_K^{\,n} = 0$.

Theorems of this type are proved in a very interesting paper by B. Hartley, The residual nilpotence of wreath products, Proc. London Math. Soc. (to appear in 1970).

4.4 Power Series.

The older method of studying free group rings, due to Magnus [7], is by power series. (Cf. also [9].) We consider the connexion between the two methods.

Let $Y = (y_i)_{i \in I}$ be a given family of elements and form the free associative ring A on Y. This is a graded ring, the piece A_n of degree n being the additive group generated by all Y-monomials of degree n.

If $\overline{A}_n$ is the ideal generated by A_n, then

$$P = \varprojlim A/\overline{A}_n$$

is the ring of all formal power series in Y: $P = \prod_{n=0}^{\infty} A_n$. The natural homomorphism $A \to P$ is an injection.

PROPOSITION 4. *The mapping $\delta: \mathbb{Z}F \to P$ given by*

$$\alpha\delta = \alpha\varepsilon + \Sigma\, y_i(\alpha d_i \varepsilon) + \Sigma\, y_i y_j(\alpha d_i d_j \varepsilon) + \dots$$

is a ring monomorphism. The image lies in the subring of all power series that involve only a finite number of the variables.

Proof. Clearly δ is additive and it is one-one by Theorem 1. There remains the multiplicative property:

$$(\alpha\beta)d_i = (\alpha d_i)\beta + (\alpha\varepsilon)(\beta d_i) ;$$

and so (induction on r):

$$(\alpha\beta)d_{i_1}\ldots d_{i_r}\varepsilon = \sum_{s=0}^{r} (\alpha d_{i_1}\ldots d_{i_s}\varepsilon)(\beta d_{i_{s+1}}\ldots d_{i_r}\varepsilon).$$

Hence δ is multiplicative.

For any ring Λ, let $U(\Lambda)$ denote the group of invertible elements ("group of units") in Λ. ($U(\cdot)$ is a functor on the category of rings because all our rings have identity elements.) By Proposition 4, $U(P)$ contains a copy of F. Note that $x_i\delta = 1+y_i$.

PROPOSITION 5. $U(P) = \{p(Y) \mid p(0) = \pm 1\}$.

Proof. If $p\in U(P)$, $p(0)$ is an invertible element in $\mathbb{Z}$.

Conversely, if $p = p_0+p_1+\ldots$, $q = q_0+q_1+\ldots$ and $pq = 1$,

then
$$q_0 = p_0 ;$$
$$p_1 + q_1 = 0 : \text{ solve for } q_1;$$
$$p_2q_0 + p_1q_1 + p_0q_2 = 0 : \text{ solve for } q_2 ;$$
etc.,

thus giving q.

Let P_n be the ideal in P generated by A_n: $P_n = \prod_{i=n}^{\infty} A_i$.
If $U_{(n)} = (1+P_n)\cap U(P)$, $n \geq 1$, then $(U(P):U_{(1)}) = 2$ (Proposition 5) and $U_{(n)} \geq U(P)_n$ for all $n > 1$. ($U(P)_n$ = n-th lower central term.)
Moreover,
$$U(P) \cong \varprojlim U(P)/U_{(n)}$$

and the embedding $F \to U_{(1)}$ induces embeddings $F/F_n \to U_{(1)}/U_{(n)}$ for all $n \geq 1$: for $\mathfrak{f}^n \delta = P_n \cap (\mathbb{Z}F)\delta$ (by Proposition 4 and Lemma 1) and hence $F \cap U_{(n)} = F_n$ (by Magnus' theorem).

4.5 Units and zero divisors.

The group U(P) of the last section is enormous. By contrast, we now show that $U(\mathbb{Z}F)$ is as small as it possibly can be.

Let $\mathfrak{X}$ be the class of all groups G such that every finitely generated subgroup $H \neq 1$ possesses a normal subgroup K so that H/K is infinite cyclic.

Clearly, $\mathfrak{X}$ includes all free groups. More generally, it is not difficult to see that $\mathfrak{X}$ includes all free products of $\mathfrak{X}$ -groups. Moreover, $\mathfrak{X}$ is extension closed (i.e., if G/N and $N \in \mathfrak{X}$ then $G \in \mathfrak{X}$), residually closed (i.e., $R\mathfrak{X} = \mathfrak{X}$) and (obviously) subgroup closed.

THEOREM 3. (G. Higman [3].) If $G \in \mathfrak{X}$, then

(i) $\mathbb{Z}G$ has no zero divisors;

(ii) $U(\mathbb{Z}G) = U(\mathbb{Z}) \times G$.

(A similar result holds with any integral domain as coefficient ring.)

Proof. Take $\alpha, \beta \neq 0$ in $\mathbb{Z}G$. We must prove that (i) $\alpha\beta \neq 0$ and (ii) if $\pm\alpha \notin G$ or $\pm\beta \notin G$, then $\alpha\beta \notin G$.

Without loss of generality, 1 occurs in both α and β. Let H be the subgroup generated by the n_α elements of G in α and the n_β elements of G in β.

<u>(i)</u>: Induction on $n = n_\alpha + n_\beta$. If $n = 2$, we are done. If $n > 2$, $H \neq 1$ and so there exists $K \lhd H$ so that H/K is infinite cyclic. If aK is a generator, then $A = \langle a \rangle$ is a transversal of K in H. We view A as an ordered group in the natural way. We may write

$$\alpha = \alpha_1 u_1 + \ldots + \alpha_r u_r, \text{ where } u_1 < \ldots < u_r \text{ in } A,\ \alpha_i \in \mathbb{Z}K,$$

$$\beta = \beta_1 v_1 + \ldots + \beta_s v_s, \text{ where } v_1 < \ldots < v_s \text{ in } A,\ \beta_i \in \mathbb{Z}K.$$

Since 1 occurs in α, $r = 1$ implies $\alpha \in \mathbb{Z}K$. If $r = s = 1$, then α and β lie in $\mathbb{Z}K$: but this is impossible as $K \neq H$. So <u>$r > 1$ or $s > 1$</u>. Let

$$\alpha\beta = \gamma_1 w_1 + \ldots + \gamma_m w_m, \qquad w_1 < \ldots < w_m \text{ in } A,\ \gamma_i \in \mathbb{Z}K,$$

Since $\min\{u_i v_j\} = u_1 v_1$ and all other $u_i v_j$ strictly exceed $u_1 v_1$, $\gamma_1 w_1 = (\alpha_1 u_1)(\beta_1 v_1)$; and since $\max\{u_i v_j\} = u_r v_s$ and all other $u_i v_j$ are strictly smaller than $u_r v_s$, $\gamma_m w_m = (\alpha_r u_r)(\beta_s v_s)$.

The induction on n yields $(\alpha_1 u_1)(\beta_1 v_1) \neq 0$ (because $r > 1$ or $s > 1$). But $\alpha\beta = 0$ implies $\gamma_1 w_1 = 0$. So $\alpha\beta \neq 0$.

<u>(ii)</u>: By hypothesis, $r > 1$ or $s > 1$. Hence $(\alpha_1 u_1)(\beta_1 v_1) \neq 0$ and $(\alpha_r u_r)(\beta_s v_s) \neq 0$ by (i). Thus $\gamma_1 w_1 \neq 0$ and $\gamma_m w_m \neq 0$ in $\alpha\beta$, so that $\alpha\beta \notin G$.

<u>Problem</u>. What does Theorem 3 become for torsion-free groups?

Sources and references.

The relation between free groups, free associative rings and free Lie rings is made particularly transparent in chapter 6 of P. Hall's Canadian notes. A proof of Magnus's theorem (§4.1) can easily be read off from the discussion there.

[1] Crowell, R.H. and Fox, R.H.: Introduction to knot theory, Ginn and Co., 1963.

[2] Fox, R.H.: Free differential calculus, I, Annals Math. 57 (1953) 547-560.
Warning: Disregard lines 6 to 15 of page 557.

[3] Higman, G.: The units of group rings, Proc. London Math. Soc. 46 (1940) 231-248.

[4] Jennings, S.A.: The structure of the group ring of a p-group over a modular field, Trans. Amer. Math. Soc. 50 (1941) 175-185.

[5] Jennings, S.A.: The group ring of a class of infinite nilpotent groups, Canadian J. Math. 7 (1955) 169-187.

[6] Lazard, M.: Sur les groupes nilpotents et les anneaux de Lie, Ann. École Norm. 71 (1954) 101-190.

[7] Magnus, W.: Beziehungen zwischen Gruppen und Idealen in einem speziellen Ring, Math. Ann. 111 (1935) 259-280.

[8] Magnus, W.: Über Beziehungen zwischen höheren Kommutatoren, Crelle 177 (1937), 105-115.

[9] Magnus, W., Karrass, A. and Solitar, D.: Combinatorial group theory, Interscience, 1966.

[10] Mal'cev, A.I.: Generalized nilpotent algebras and their associated groups, Mat.Sbornik N.S. 25 (67) (1949) 347-366. Amer. Math. Soc. Transl. (2) 69 (1968).

[11] Schumann, H.G.: Über Moduln und Gruppenbilder, Math.Ann. 114 (1935) 385-413.

[12] Zassenhaus, H.: Ein Verfahren, jeder endlichen p-Gruppe einen Lie-Ring zuzuordnen, Abh. math. Sem. Hamburg, 13 (1940) 200-207.

[13] Hoare, A.H.M.: Group rings and lower central series, J. London Math. Soc. (2) 1 (1969) 37-40.

[14] Passi, I.B.S.: Dimension subgroups, J. of Algebra 9 (1968) 152-182.

[6] is a mine of information concerning all the topics discussed in this chapter, except §4.5 Cf. also Lazard's newer paper: Groupes analytiques p-adiques, I.H.E.S., no. 26, 1965.

[10] appeared only on the side lines of this chapter. It is therefore worth stressing that this is an exceedingly interesting paper containing many ideas that have still not been followed up as much as they deserve.

CHAPTER 5

CLASSICAL EXTENSION THEORY

5.1 The problem.

For a given group G, we wish to consider all extensions by G. These form a category in a natural way: the objects are all exact sequences

$$(*):\ 1 \to K \to E \xrightarrow{\pi} G \to 1\ ;$$

and if

$$(*'):\ 1 \to K' \to E' \xrightarrow{\pi'} G \to 1$$

is another extension, then a morphism $(*) \to (*')$ is a pair (α, ρ) of group homomorphisms so that

$$\begin{array}{ccccccccc} 1 & \to & K & \to & E & \to & G & \to & 1 \\ & & \downarrow{\scriptstyle\alpha} & & \downarrow{\scriptstyle\rho} & & \downarrow{\scriptstyle=} & & \\ 1 & \to & K' & \to & E' & \to & G & \to & 1 \end{array}$$

commutes. Denote this category by $\left(\frac{G}{\;}\right)$.

Consider a particular extension (*). Let In K be the group of inner automorphisms of K and Out K = Aut K / In K . Then there exists χ completing the square

$$\begin{array}{ccc} E & \xrightarrow{\text{conjugation}} & \text{Aut } K \\ {\scriptstyle\pi}\downarrow & & \downarrow \\ G & \xrightarrow{\quad\chi\quad} & \text{Out } K \end{array} .$$

We call χ the <u>coupling</u> of G to K determined by (*).

Remarks. (1) If K is abelian, χ defines a G-module structure on K.

(2) More generally, if Z is the centre of K, then the restriction: Aut K → Aut Z gives a homomorphism β: Out K → Aut Z and $\chi\beta$ induces a G-module structure on Z.

It is sensible, as a first step, to consider only extensions by G over a fixed group K with a fixed coupling K. Write $\left(\frac{G}{K,\chi}\right)$ for the full subcategory of all these.

Definition. (*), (*') are called equivalent if there exists a morphism $(1,\rho)$: (*) → (*').

Let $\left[\frac{G}{K,\chi}\right]$ be the class (actually a set, as we shall see) whose objects are the equivalence classes of extensions with the above meaning of equivalence.

Classical extension theory consists of three parts:

(1) A description of $\left[\frac{G}{K,\chi}\right]$ when K is abelian (§5.3, below).

(2) A description of $\left[\frac{G}{K,\chi}\right]$ in terms of $\left[\frac{G}{Z,\chi'}\right]$, where $Z = \zeta_1(K)$, the centre of K, and χ' is G → Aut Z via χ (§5.4, below).

(3) A study of when a given homomorphism χ : G → Out K provides a non-empty $\left[\frac{G}{K,\chi}\right]$ (§5.5, below).

Notation. When K is abelian, we shall write $\left(\frac{G}{K}\right)$ instead of $\left(\frac{G}{K,\chi}\right)$, where K in the former symbol is viewed as a G-module.

Similary, $\left[\frac{G}{K}\right]$ will stand for $\left[\frac{G}{K,\chi}\right]$.

5.2 Covering groups

Let U, K be given groups and ξ a homomorphism of U into Aut K. Using ξ, we may construct the split extension S = U[K. Suppose V is a normal subgroup of U such that

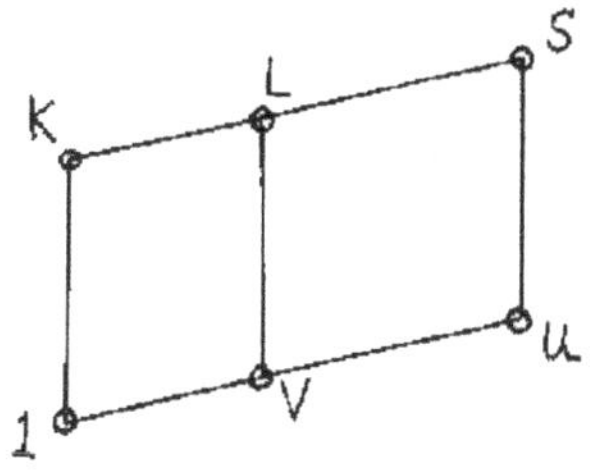

$$[V,Z] = 1, \tag{1}$$

where $Z = \zeta_1(K)$, the centre of K. Put L = VK.

We shall consider the set $\mathfrak{M}$ of all normal subgroups M of S such that L = M x K. (Note that $V \in \mathfrak{M}$ if, and only if, V centralizes K: in view of (1), this is automatic when K is abelian.) Every element M of $\mathfrak{M}$ provides us with an extension

$$1 \to K \to S/M \to U/V \to 1 \tag{2}$$

and, conversely, if (2) is given (with the natural homomorphisms) then $M \in \mathfrak{M}$. It is quite possible for $\mathfrak{M}$ to be empty. We shall assume for the remainder of this section that this is not the case.

LEMMA 1. *If $\varphi \in \mathrm{Hom}_U(V,Z)$ and φ' is the mapping*

$$vk \to (vk)(v\varphi),$$

then $\varphi' \in \mathrm{Aut}_U(L)$ and

$$\varphi'' : M \to M\varphi'$$

defines a permutation of $\mathfrak{M}$. Moreover, $\varphi \to \varphi''$ is a regular permutational representation of $\mathrm{Hom}_U(V,Z)$ on $\mathfrak{M}$.

Proof. All is routine except perhaps that the representation is regular. Suppose $M\varphi' = M$. If $\varphi' \neq 1$, there exists v so that $v\varphi \neq 1$. Write $v = v_M k$, where $v_M \in M$, $k \in K$. So $v_M\varphi' \in M\varphi' = M$. But

$$v_M\varphi' = v\ k^{-1}\ (v\varphi),$$

whence $v\varphi \in M$. So $M \cap K \neq 1$, a contradiction. Thus $\varphi \to \varphi''$ is semi-regular.

To see transitivity, let $M, N \in \mathfrak{M}$ and $v \in V$. If

$$v = v_M k = v_N k',$$

where $v_M \in M$, $v_N \in N$ and $k, k' \in K$, then k and k' induce the same automorphism on K. Hence $k'(v\varphi) = k$, for some $v\varphi$ in Z. Then it is easy to see that $\varphi \in \mathrm{Hom}_U(V,Z)$ and we assert $M\varphi' = N$. For if $m \in M$, write $m = vx$ with v in V and x in K and $v = v_M k = v_N k'$, as above. So $kx \in M \cap K = 1$. Now

$$m\varphi' = (vx)(v\varphi) = v_N k'\ (v\varphi)x = v_N kx = v_N,$$

so that $M\varphi' \leq N$. But $M\varphi' \in \mathfrak{M}$ and so $M\varphi' = N$.

COROLLARY. There is a one-one correspondence between $\mathfrak{M}$ and $\mathrm{Hom}_U(V,Z)$.

We define an equivalence relation on $\mathfrak{M}$ by requiring M, N in $\mathfrak{M}$ to be equivalent, $M \sim N$, whenever the corresponding extensions (cf. (2)) are equivalent: i.e., if there exists a homomorphism θ such that

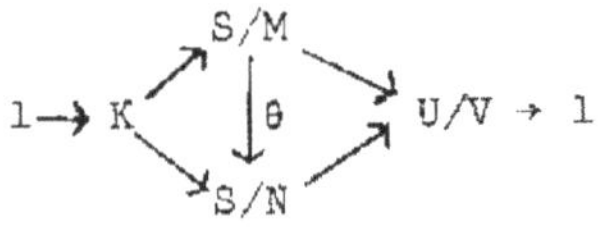

commutes.

Suppose this is so. Clearly, $(kM)\theta = kN$ for all k in K. For any u in U, uM, uN both map to uV and so $(uM)\theta$, uN differ by an element of L/N:

$$(uM)\theta = ucN,$$

where $c \in K$ and c is uniquely determined by u (since $K \cap N = 1$). We assert $c \in Z$. For if $a \in K$, $(a^u M)\theta = a^u N$, while

$$(u^{-1}auM)\theta = (uc)^{-1}a(uc)N,$$

because θ is a homomorphism, so that $a^u = a^{uc}$. Since this holds for all a in K, c is central in K. Thus $u \to c$ is a mapping, call it δ, of U into Z. As θ is a homomorphism, $\delta \in \mathrm{Der}(U,Z)$.

The inclusion $V \to U$ gives the usual restriction mapping

$$\mathrm{Der}(U,Z) \to \mathrm{Hom}_U(V,Z)$$

and this enables us to view Der(U,Z) as acting on $\mathfrak{M}$. If δ is the derivation considered above, we may construct δ' as in Lemma 1 and clearly $(xM)\theta = (x\delta')N$ for x in L. Hence $M\delta' = N$. We have now proved the non-trivial half of

Lemma 2. If Der(U,Z) is allowed to act on $\mathfrak{M}$ via the restriction $\mathrm{Der}(U,Z) \to \mathrm{Hom}_U(V,Z)$ and the use of Lemma 1, then for all M, N in $\mathfrak{M}$, we have $M \sim N$ if, and only if, M, N belong to a Der(U,Z)-orbit (transitivity class).

We sum up our conclusions in

PROPOSITION 1. Assume the set $\mathfrak{M}$ is non-empty. Then the group $\mathrm{Hom}_U(V,Z)$ acts as a regular permutation group on $\mathfrak{M}$. If Der(U,Z)

<u>is allowed to act on $\mathfrak{M}$ via the restriction mapping, then the Der(U,Z)-orbits are precisely the equivalence classes with respect to the relation $\sim$ on $\mathfrak{M}$.</u>

5.3 Extensions with abelian kernel

Let G be a given group and A a given G-module. Choose any free presentation

$$1 \to R \to F \xrightarrow{\pi} G \to 1$$

of G. If $\xi = \pi\chi$, where χ is the given homomorphism of G into Aut A, then we may form S as in §5.2, using F for U, A for K and R for V. In this case $R \in \mathfrak{M}$ and the simplest one-one correspondence between $\mathfrak{M}$ and $\mathrm{Hom}_F(R,A)$ is the one obtained by making R correspond to the zero homomorphism. (We write $\mathrm{Hom}_F(R,A)$ additively.) Then $R\varphi'$ corresponds to φ (cf. Lemma 1). Note further that the extension corresponding to R (cf. equation (2) in §5.2) is the split extension.

By MacLane's result (Proposition 6, §3.6),

$$H^2(G,A) \cong \mathrm{Coker}\,(\mathrm{Der}(F,A) \to \mathrm{Hom}_G(R/R',A)\,).$$

But $\mathrm{Hom}_F(R,A)$ is essentially the same as $\mathrm{Hom}_G(R/R',A)$. Hence, by Proposition 1, $H^2(G,A)$ is in one-one correspondence with the equivalence classes on $\mathfrak{M}$.

It remains to be seen that every extension of A by G is equivalent to one determined by an element of $\mathfrak{M}$ So consider

$$\begin{array}{ccccc} 1 \to A \to E & & & & \\ & \searrow^{\tau} & & & \\ & & G \to 1 & . & \\ & \nearrow_{\pi} & & & \\ 1 \to R \to F & & & & \end{array}$$

Since F is free, we may lift π to θ: $F \to E$ so that $\theta\tau = \pi$. Then $R\theta \leq A$ and we obtain a homomorphism θ^*: $wa \to w^{\theta}a$ of S onto E. Clearly Ker $\theta^* \epsilon$ $\mathfrak{M}$.

Note that

$$\text{Ker } \theta^* = \{r(r\theta)^{-1};\ r\epsilon R\} = R(-\theta)' \quad ,$$

in the notation of Lemma 1. For $w^{\theta}a = 1$ implies $w^{\theta} \epsilon A$ and so $w^{\theta\tau} = w^{\pi} = 1$, i.e., $w \epsilon R$. Thus

$$\text{Ker } \theta^* = \{ra \mid r\epsilon R \text{ and } a = (r\theta)^{-1}\} = \{r(r\theta)^{-1};\ r\epsilon R\}.$$

If the standard presentation of G is used (§3.4), then the restriction of θ to R is a classical factor-set of the extension E.

We have now proved

THEOREM 1. Let $\alpha \epsilon H^2(G,A)$ and θ be an element in $\mathrm{Hom}_F(R,A)$ representing α. Denote by $[\alpha]$ the class of the extension

$$1 \to A \to S/\{r(r\theta)^{-1};\ r\epsilon R\} \to G \to 1.$$

Then $\alpha \to [\alpha]$ is a one-one mapping of $H^2(G,A)$ onto the set of objects of $\left[\frac{G}{A}\right]$. Moreover, [0] is the split extension.

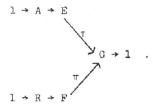

Exercise. Prove that if θ_i corresponds to M_i in $\mathfrak{M}$ and $\theta_1+\theta_2$ to M, then

$$1 \to A \to S/M \to G \to 1$$

is equivalent to

$$1 \to A \to E \to G \to 1,$$

with $E = I/J$, where I is the subgroup of $S/M_1 \times S/M_2$ of all pairs (x,y) so that x and y have the same image in G, and J is all pairs $(aM_1, a^{-1}M_2)$, $a \in A$.

Remark. How does Theorem 1 depend on the particular free presentation chosen for G?

$$\text{Let} \quad (P_i) : 1 \to R_i \to F_i \xrightarrow{\pi_i} G \to 1$$

be two free presentations of G and (R_i) the corresponding resolutions of Theorem 2, §3.1. If $\psi : F_1 \to F_2$ is a homomorphism lifting π_1, then ψ yields a morphism $(R_1) \to (R_2)$ and this, in turn, gives an isomorphism

$$\psi_A^*: \text{homology}(\text{Hom}_G((R_2),A) \to \text{homology}(\text{Hom}_G((R_1),A))$$

for each G-module A (cf. Cartan-Eilenberg, p. 77, last paragraph and p. 82, last paragraph). In fact, ψ^* is a natural isomorphism (natural equivalence) of functors. Since $H^*(G, \ )$ was defined only to within natural isomorphism, ψ^* may be regarded as the identity isomorphism on $H^*(G, \ )$.

In particular, we have the commutative triangle

$$\begin{array}{ccc} \mathrm{Hom}_G(R_2/R_2', A) & \xrightarrow{\text{via } \psi} & \mathrm{Hom}_G(R_1/R_1', A) \\ & \searrow \quad \swarrow & \\ & H^2(G,A) & \end{array} .$$

If $\alpha \epsilon H^2(G,A)$ and $\theta \searrow \alpha \swarrow \varphi$ in the above triangle, then $\psi\theta$, φ are cohomologous and hence the corresponding extensions given by Theorem 1 are equivalent. We wish to satisfy ourselves that the extension S_2/M_2 given by θ is equivalent to S_1/M_1 given by $\psi\theta$ (where $S_i = F_i[A)$. Now

$$M_2 = \{ r(r\theta)^{-1};\ r\epsilon R_2\} ,$$

$$M_1 = \{ r(r\psi\theta)^{-1};\ r\epsilon R_1\}$$

and $M_1\psi \leq M_2$. The homomorphism $S_1/M_1 \to S_2/M_2$ induced by ψ gives the required equivalence of extensions.

5.4 General extensions

Definition. Let A be a given G-module. Suppose K is a group, i an embedding of A in K so that Ai is the centre of K, and χ is a homomorphism of G into Out K such that χ induces the given G-module structure on Ai. We shall call (i,K,χ) a (G:A)-core.

Recall from §5.1 that every extension of a group K by G gives rise to a (G:A)-core, where $A = \zeta_1(K)$ and i is inclusion.

If (i,K,X) is a (G:A)-core, then we can always construct a commutative picture

$$\begin{array}{ccccccc} K & \to & \mathrm{Aut}K & \to & \mathrm{Out}K & \to & 1 \\ \uparrow{\scriptstyle\eta} & & \uparrow{\scriptstyle\xi} & & \uparrow{\scriptstyle\chi} & & \\ R & \to & F & \xrightarrow{\pi} & G & \to & 1 \end{array} \tag{1}$$

where ξ is any homomorphism lifting $\pi\chi$ and η is any homomorphism lifting the restriction of ξ to R.

Form the split extension S of K by F, using ξ. We may now apply the theory of §5.2 with F for U, R for V and Ai for Z. If $\mathfrak{M}$ is non-empty, then, by Proposition 1, the equivalence classes of $\mathfrak{M}$ are in one-one correspondence with $\mathrm{Coker}(\mathrm{Der}(F,A) \to \mathrm{Hom}_F(R,A))$, i.e., with $H^2(G,A)$.

On the other hand, the equivalence classes of $\mathfrak{M}$ are in one-one correspondence with the elements of $\left[\frac{G}{K,\chi}\right]$: Let

$$1 \to K \to E \xrightarrow{\tau} G \to 1 \tag{2}$$

be an extension with coupling χ. Lift π to γ: $F \to E$. If $e \to e'$ is the homomorphism $E \to \mathrm{Aut}\ K$, $(w\gamma)'$, $w\xi$ lie in $w\pi\chi$ and so

$$w\xi = (w\gamma)'\ k'$$

for some k in K. If F is free on X, choose, for each x in X, k_x in K so that

$$x\xi = (x\gamma)'\ k_x'\ .$$

Let $x \to (x\gamma)k_x$ extend to a homomorphism $\sigma : F \to E$. Then $\sigma\tau = \pi$ and σ induces ξ (i.e., $(w\sigma)' = w\xi$).

We now obtain a homomorphism σ^* of S onto E by setting

$$(wk)\sigma^* = (w\sigma)k \quad (w\epsilon F,\ k\epsilon K).$$

The kernel of σ^* belongs to $\mathfrak{M}$ and the extension corresponding to it, i.e.,

$$1 \to K \to S/\mathrm{Ker}\ \sigma^* \to G \to 1$$

is equivalent to (2) above.

(Note that, if the standard presentation is used, then the restriction of σ to R is a classical factor-set for the extension (2).)

We have now proved the first part of

THEOREM 2. *Let (i,K,χ) be a given (G:A)-core.*

(i) If $\left[\frac{G}{K,\chi}\right]$ is non-empty, then there is a one-one correspondence between $\left[\frac{G}{K,\chi}\right]$ and $H^2(G,A)$.

(ii) If ξ,η are homomorphisms as in (1) above and also satisfying

$$(r^w)\eta = (r\eta)^{w\xi}$$

for all r in R, w in F, then there exists an extension

$$1 \to K \to E \xrightarrow{\tau} G \to 1$$

and a homomorphism σ: F → E so that $\sigma\tau = \pi$, σ induces ξ and σ restricted to R is η.

Part (ii) is immediate if we set M to be $\{r(r\eta)^{-1};\ r\epsilon R\}$ in our split extension S. For then $M \in \mathfrak{M}$,

$$1 \to K \to S/M \to G \to 1$$

is the required extension and $w \to wM$ the required homomorphism.

Remarks.

(1) If $\zeta_1(K) = 1$, $\left[\frac{G}{K,\chi}\right] \neq \phi$: for then $\xi|R$ can be taken for η in part (ii). Of course we have then card $\left[\frac{G}{K,\chi}\right] = 1$ by part (i).

We may see this directly: The "pull-back" for

$$\begin{array}{ccccccccc} & & & & & & G & & \\ & & & & & & \downarrow & & \\ 1 & \to & K & \to & \text{Aut } K & \to & \text{Out } K & \to & 1 \end{array}$$

gives an extension of K by G with respect to χ and it is unique.

(2) The correspondence in part (i) is not natural - unlike that of Theorem 1: for here $\mathfrak{M}$ has no distinguished element.

(3) card $\left[\frac{G}{K,\chi}\right] = 1$ gives a small amount of negative information but none about the extension class that does exist.

Example: if G is not free, F is not cyclic and $1 \to R \to F \to G \to 1$ is exact with coupling χ say, then card $\left[\frac{G}{K,\chi}\right] = 1$ (because $\zeta_1(R) = 1$), but the extension is not split. Hence a split extension of R by G cannot exist.

5.5 Obstructions

Again take a free presentation

$$1 \to R \to F \to G \to 1$$

of G and let A be a given G-module. If (i,K,χ) is a $(G{:}A)$-core, construct ξ, η as in (1) of §5.4. For each r in R, w in F, the elements $r\eta$, $\left(r^{w^{-1}}\eta\right)^{w\xi}$ induce the same inner automorphism of K. Hence

$$w \circ r = (r^{-1}\eta)\left(r^{w^{-1}}\eta\right)^{w\xi}$$

is an element of Ai. We propose henceforth to identify A and Ai. It is a simple matter to check the following two rules:

$$w \circ r_1 r_2 = (w \circ r_1)(w \circ r_2), \tag{1}$$

for all w in F, r_1, r_2 in R (because A is central in K); and

$$w_1 w_2 \circ r = (w_2 \circ r)\left(w_1 \circ r^{w_2^{-1}}\right)^{w_2\xi}, \tag{2}$$

for all w_1, w_2 in F, r in R.

We consider now the resolution determined functorially by our presentation of G (§3.1). Let $\{x_i;\ i \in I\}$ be a set of free generators of F and $\{y_j\ ;\ j \in J\}$ a set of free generators of R. Since $\mathfrak{f}\mathfrak{r}/\mathfrak{f}\mathfrak{r}^2$ is free as G-module on the set of cosets of the elements $(1-x_i)(1-y_j)$, for all i in I, j in J, the mapping

$$(1-x_i)(1-y_j) + \mathfrak{f}\mathfrak{r}^2 \to x_i \circ y_j$$

extends to a G-module homomorphism φ of $\mathfrak{f}\mathfrak{r}/\mathfrak{f}\mathfrak{r}^2$ into A.

Equation (1) implies that

$$((1-x_i)(1-r) + \mathfrak{f}\mathfrak{r}^2)\varphi = x_i \circ r$$

for all r in R; and an induction on the length of w together with (2) and the identity

$$(1-ab)(1-r) = (1-b)(1-r) + (1-a)\left(1-r^{b^{-1}}\right)b$$

shows that

$$((1-w)(1-r) + \mathfrak{f}\mathfrak{r}^2)\varphi = w \circ r$$

for all w in F (and all r in R). It follows that φ restricts to zero on $\mathfrak{r}^2/\mathfrak{f}\mathfrak{r}^2$. Hence φ yields an element of $\mathrm{Hom}_G(\mathfrak{f}\mathfrak{r}/\mathfrak{r}^2, A)$. We shall also denote this element by φ and we call it the <u>obstruction</u> determined by ξ, η.

Our resolution and Proposition 1, §2.1 show that

$$H^3(G,A) \cong \mathrm{Coker}\ (\mathrm{Hom}_G(\mathfrak{r}/\mathfrak{r}^2, A) \to \mathrm{Hom}_G(\mathfrak{f}\mathfrak{r}/\mathfrak{r}^2, A)).$$

Consequently φ determines a 3-dimensional cohomology class.

<u>LEMMA 3. The 3-dimensional cohomology class determined by φ is independent of the choice of ξ and η.</u>

<u>Preparatory remark</u>. The definition of derivation given in §3.1 and Lemma 2 there both apply, with no change, to G-groups instead of G-modules. Consequently, any mapping of $\{x_i\}$ into an F-group K extends to a derivation of F in K.

<u>Proof</u>. Let η' be another homomorphism lifting ξ. Then, for all r in R,

$$r\eta' = (r\eta)\ a_r$$

for some a_r in A. If ψ is the G-module homomorphism $\mathfrak{r}/\mathfrak{r}^2 \to A$ extending $(1-y_j) + \mathfrak{r}^2 \to a_{y_j}$, then

$$\psi\colon (1-r) + \mathfrak{r}^2 \to a_r$$

for all r in R. If ξ,η' yield the obstruction φ', then $\varphi' = \varphi-\psi$, because

$$(1-w)(1-r) + \mathfrak{r}^2 = (1-r) - \left(1-r^{w^{-1}}\right)w + \mathfrak{r}^2 \xrightarrow{-\psi} a_{r^{-1}}\left(a_{r^{w^{-1}}}\right)^{w\xi} .$$

Note that, by varying η', we obtain all ψ in $\mathrm{Hom}_G(\mathfrak{r}/\mathfrak{r}^2, A)$.

Next, let ξ' be another homomorphism lifting $\pi\chi$. For each i in I we can choose k_i in K so that

$$x_i\xi' = x_i\xi\ k_i'$$

(where k' denotes the inner automorphism determined by k). Viewing K as an F-group via ξ, we extend $x_i \to k_i$ to a derivation δ of F in K and then

$$w\xi' = w\xi(w\delta)' \tag{3}$$

for all w in F (preparatory remark, above). If we define $\eta'\colon R \to K$ by

$$r\eta' = (r\eta)(r\delta),$$

then clearly η' is a homomorphism and lifts ξ' (by (3)). We now check that ξ',η' yield the same obstruction φ as ξ,η:

$$(r^{-1}\eta')\left(r^{w^{-1}}\eta'\right)^{w\xi'} = (r\eta\ r\delta)^{-1}\left(r^{w^{-1}}\eta\ r^{w^{-1}}\delta\right)^{w\xi}\ w\delta \tag{4}$$

and, since $\left(r^{w^{-1}}\delta\right)^{w\xi} = (wr)\delta\ (w\delta)^{-1}$, the right hand side of (4) is

$$\begin{aligned}(r\delta)^{-1}\ (r\eta)^{-1}\left(r^{w^{-1}}\eta\right)^{w\xi}\ w\delta\ (w\delta)^{-1}\ (wr)\delta \\ = (r\delta)^{-1}\ (w\delta)^{-r\eta}(w\circ r)\ (wr)\delta \\ = w\circ r.\end{aligned}$$

<u>PROPOSITION 2. $\left[\frac{G}{K,\chi}\right]$ is non-empty if, and only if, (i,K,χ) determines the zero element of $H^3(G,A)$.</u>

<u>Proof</u>. If $1 \to K \to E \xrightarrow{\tau} G \to 1$ is an extension and $\sigma\colon F \to E$ is a homomorphism lifting π, let ξ be the composite $F \to E \to \mathrm{Aut}\,K$ and η the restriction of σ to R. Then ξ,η determine the zero homomorphism of $\mathfrak{f}\mathfrak{r}/\mathfrak{r}^2$ into A.

Conversely, if (i,K,χ) determines the zero element in $H^3(G,A)$, we know that we can choose ξ,η so that the corresponding obstruction: $\mathfrak{f}\mathfrak{r}/\mathfrak{r}^2 \to A$ is zero. Thus $(r^w\eta) = (r\eta)^{w\xi}$, for all r in R, w in F, and hence the result follows from Theorem 2 (ii).

Two cores (i,K,χ), (i', K', χ') shall be called <u>similar</u> if they determine the same element of $H^3(G,A)$. The set of similarity classes of (G:A)-cores will be written $\left[\frac{G}{A}\right]$. This set has as distinguished element the class containing the core (i,A,χ), where i is the identity on A and χ is the homomorphism $G \to \mathrm{Aut}\,A$ giving the G-module structure of A.

<u>THEOREM 3. There is a one-one correspondence between $\left[\frac{G}{A}\right]$ and $H^3(G,A)$ in which the similarity class containing (i,A,χ) corresponds to 0.</u>

It only remains to be proved that the mapping $\left[\frac{G}{A}\right] \to H^3(G,A)$ is surjective.

Let us assume our presentation of G is big enough so that R is non-cyclic. Then $R \times A$ has centre A. (If the standard presentation is used, then R is cyclic if, and only if, G is of order 2.

This explains the curious case distinction necessary at this stage in the older treatments.)

Take any φ in $\mathrm{Hom}(\mathfrak{f}\mathfrak{r}/\mathfrak{r}^2, A)$ and for each w in F let $w\xi$ be the mapping

$$(r,a) \to (r^w,\ a(w\pi) + ((1-w)(1-r^w) + \mathfrak{r}^2)\varphi).$$

It is completely straightforward to check that $w\xi$ is an automorphism of $R \times A$ and then that ξ is a homomorphism of F into $\mathrm{Aut}(R \times A)$. Clearly, for each r in R, $r\xi$ is conjugation by $(r,0)$. Thus ξ induces a homomorphism $\chi\colon G \to \mathrm{Out}(R \times A)$ and so $(1, R \times A, \chi)$ is a (G:A)-core. If η is $r \to (r,0)$, then the original φ is precisely the obstruction determined by ξ,η: for

$$\left(r^{w^{-1}}\eta\right)^{w\xi} = \left(r^{w^{-1}}, 0\right)^{w\xi} = (r,\ ((1-w)(1-r) + \mathfrak{r}^2)\varphi)$$

$$= (r,0)(1,\ ((1-w)(1-r) + \mathfrak{r}^2)\varphi).$$

This completes the proof of Theorem 3.

<u>Remarks.</u>

<u>(1)</u> : The mapping $\varphi \to \chi$ constructed above is a one-one mapping of $\mathrm{Hom}_G(\mathfrak{f}\mathfrak{r}/\mathfrak{r}^2, A)$ onto the set of all those homomorphisms $\chi\colon G \to \mathrm{Out}(R \times A)$ consistent with the G-module structure on A and such that

$$\begin{array}{ccc} & & \mathrm{Out}\,(R \times A) \\ & \overset{\chi}{\nearrow} & \downarrow \\ G & \underset{\theta}{\searrow} & \mathrm{Out}\ R \end{array}$$

commutes, where θ is the coupling of G to R determined by our free presentation of G.

Clearly, $\varphi \to \chi$ is one-one. To see surjectivity, take χ and lift to $\xi\colon F \to \mathrm{Aut}(R \times A)$ so that

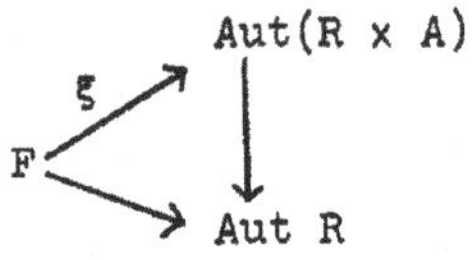

commutes. Then

$$(r,a)^{w\xi} = (r^{w}, a(w\pi) + f(w,r)),$$

for some function f. Take $\eta\colon r \to (r,1)$. Then

$$w \circ r = f\left(w, r^{w^{-1}}\right)$$

and

$$\varphi\colon (1-w)(1-r) + \mathfrak{r}^{2} \to w \circ r$$

yields χ.

(2) : Given $[i', K', \chi']$, $[i'', K'', \chi'']$ in $\left[\frac{G}{A}\right]$, let K be the central product of K' and K'':

$$K = K' \times K'' / \{ (ai', (ai'')^{-1});\ a \in A\}.$$

Write γ for the natural homomorphism $K' \times K'' \to K$. Let $\xi\colon F \to \mathrm{Aut}\,K$ be

$$w\xi : (k', k'')\gamma \to ((k')^{w\xi'}, (k'')^{w\xi''})\gamma ,$$

where ξ', ξ'' have obvious meanings; and let χ be $G \to \mathrm{Out}\,K$ given by ξ. Then, in the correspondence given by Theorem 3, $[i,K,\chi]$ corresponds to the sum of the cohomology classes determined by $[i', K', \chi']$, $[i'', K'', \chi'']$.

<u>(3) :</u> Let $(P_i): \quad 1 \to R_i \to F_i \xrightarrow{\pi_i} G \to 1$

be two free presentations of G and $\psi: F_1 \to F_2$ a homomorphism lifting π_1. If φ is the obstruction determined by ξ_2, η_2 (from (P_2)), then $\tilde{\psi}\varphi$ is the obstruction determined by $\psi\xi_2$, $\psi\eta_2$ (from (P_1)), where $\tilde{\psi}$ is the homomorphism determined by ψ in the following commutative triangle:

$$\begin{array}{ccc} \mathrm{Hom}_G(\mathfrak{f}_2\mathfrak{r}_2/\mathfrak{r}_2^{\,2}, A) & \xrightarrow{\tilde{\psi}} & \mathrm{Hom}_G(\mathfrak{f}_1\mathfrak{r}_1/\mathfrak{r}_1^{\,2}, A) \\ & \searrow \quad \swarrow & \\ & H^3(G,A) & \end{array} .$$

(For the commutativity cf. the remark after Theorem 1, §5.3.) Thus φ, $\tilde{\psi}\varphi$ determine the same 3-dimensional cohomology class.

Sources and references.

The construction in §5.2 appeared in a slightly different form and different context in lectures by P. Hall in the early 1950's. A discussion of extensions with abelian kernel in the same spirit as the one given here may be found in [1].

The modern version of extension theory is due to Eilenberg-Mac Lane [2], [3]. Cf. also Mac Lane's book [5]. The treatment given here however differs substantially from this.

[1] Barr, M. and Rinehart, G.S.: Cohomology as the derived functor of derivations, Trans. Amer. Math. Soc. 122 (1966) 416-426.

[2] Eilenberg, S. and Mac Lane, S.: Cohomology theory in abstract groups I, Annals of Math. 48 (1947) 51-78.

[3] Eilenberg, S. and Mac Lane, S.: Cohomology theory in abstract groups II, Annals of Math. 48 (1947) 326-341.

[4] Gruenberg, K.W.: A new treatment of group extensions, Math. Zeit. 102 (1967) 340-350.

[5] Mac Lane, S.: Homology, Springer, 1963.

CHAPTER 6

MORE COHOMOLOGICAL MACHINERY

6.1 Natural homomorphisms of cohomological functors.

Let F_1, F_2 be (covariant) functors from Mod_G (= category of G-modules) to $\mathfrak{A}$ (= abelian groups). A <u>natural homomorphism</u> (natural transformation) $\varphi : F_1 \to F_2$ is a family of homomorphisms (φ_A), indexed by the G-modules A, so that

$$\varphi_A : AF_1 \to AF_2$$

and whenever $f : A \to B$ in Mod_G, then

$$\begin{array}{ccc} AF_1 & \xrightarrow{\varphi_A} & AF_2 \\ {\scriptstyle fF_1}\downarrow & & \downarrow{\scriptstyle fF_2} \\ BF_1 & \xrightarrow{\varphi_B} & BF_2 \end{array}$$

is commutative.

A <u>cohomological extension</u> $(C,d) = ((C^q), (d^q);\ q = 0, 1,2,\ldots)$ of an arbitrary functor $F : \text{Mod}_G \to \mathfrak{A}$ is defined in exactly the same way as was a cohomological extension of the special functor $A \mapsto A^G$ of §2.1. We call (C,d) a <u>cohomological functor</u> (connected sequence of functors). As in §2.1, we say (C,d) is <u>minimal</u> if $C^q(A) = 0$ for all $q > 0$ and all coinduced A.

If (C_1,d_1), (C_2,d_2) are cohomological functors, then a natural homomorphism (natural transformation)

$$\psi = (\psi^q) : (C_1,d_1) \to (C_2,d_2)$$

is a sequence of natural homomorphisms

$$\psi^q : C_1{}^q \to C_2{}^q \qquad (q = 0, 1, 2, \ldots),$$

so that ψ commutes with the connecting homomorphisms: i.e., if $0 \to A \to B \to C \to 0$ is exact, then

$$\begin{array}{c} \text{cohomology sequence for } C_1 \\ \Big\downarrow \psi \\ \text{cohomology sequence for } C_2 \end{array}$$

commutes.

We say ψ is a *(cohomological) extension* of φ if $\psi^0 = \varphi$.

THEOREM 1. *Let F_1, F_2 be functors: $\mathrm{Mod}_G \to \mathfrak{A}$ and φ a natural homomorphism: $F_1 \to F_2$. If (C_i, d_i) is a cohomological extension of F_i $(i = 1,2)$ and (C_1, d_1) is minimal, then there exists one and only one natural homomorphism $\psi : (C_1, d_1) \to (C_2, d_2)$ extending φ.*

Proof. Take any A and embed A in an injective G-module I:

$$0 \to A \to I \to A' \to 0 .$$

(Proof that this is possible: Embed A in A^* and embed the abelian group A in a divisible abelian group D. Then D^* is G-injective and $A \to A^* \to D^*$ is an embedding.) Since I is a direct summand of I^*, and $C_1{}^q(I^*) = 0$, $q > 0$, therefore $C_1{}^q(I) = 0$, all $q > 0$.

Then the diagram

$$\begin{array}{ccccccccc} 0 \to & AF_1 & \to & IF_1 & \to & A'F_1 & \to & C_1{}^1(A) & \to 0 \\ & \varphi_A \downarrow & & \varphi_I \downarrow & & \varphi_{A'} \downarrow & & & \\ 0 \to & AF_2 & \to & IF_2 & \to & A'F_2 & \to & C_2{}^1(A) & \to C_2{}^1(I) \end{array}$$

enables us to construct

$$\psi_A{}^1 : \; C_1{}^1(A) \to C_2{}^1(A) \quad ;$$

and clearly there is only one such homomorphism completing this. It is easy to see that ψ^1 is a natural homomorphism.

To check that ψ^1 commutes with the differentiations (connecting homomorphisms), consider

$$\begin{array}{ccccccccc} 0 & \to & A & \to & B & \to & C & \to & 0 \\ & & = \downarrow & & \beta\downarrow & & \gamma\downarrow & & \\ 0 & \to & A & \to & I & \to & A' & \to & 0 \end{array}$$

where the top row is any given short exact sequence, the lower row is as above, β exists by the injectivity of I, and γ is the mapping induced by β. We now obtain a cube

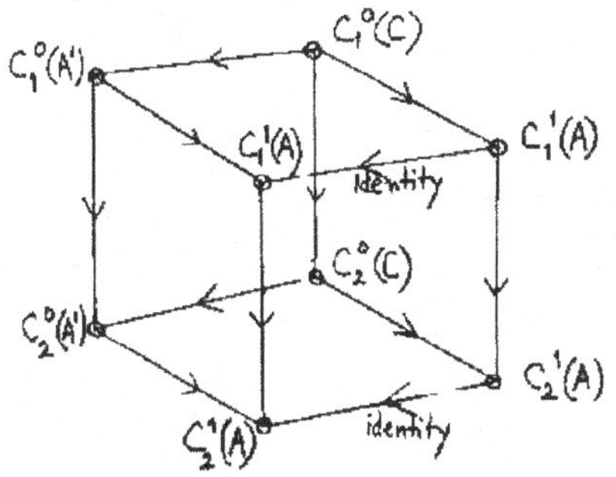

The top and bottom faces are commutative because C_1, C_2 are cohomological functors; the far face (i.e., the "A'-C face") is commutative because $\psi^0 = \varphi$ is a natural homomorphism; the "A'-A face" is commutative by the construction of ψ^1. Hence the "C-A face" is commutative.

The original sequence $0 \to A \to I \to A' \to 0$ yields

$$\begin{array}{ccccccc} & & 0 & \to C_1^{\,1}(A') & \longrightarrow & C_1^{\,2}(A) \to 0 & \\ & & & \downarrow \psi_{A'}^{\,1} & & & \\ C_2^{\,1}(I) & \to & & C_2^{\,1}(A') & \longrightarrow & C_2^{\,2}(A) \to C_2^{\,2}(I) & \end{array}$$

and this enables us to define

$$\psi_A^{\,2} : C_1^{\,2}(A) \to C_2^{\,2}(A).$$

The check that ψ^2 is a natural homomorphism commuting with the differentiations is essentially as above.

We continue in this manner, thus producing our cohomological extension ψ of φ.

6.2 Restriction, inflation, corestriction.

In this section let V be the functor $A \mapsto A^G$. We shall apply Theorem 1 in a number of special cases.

(1) Scalar multiplication.

$F_1 = F_2 = V$, $C_1 = C_2 = H^*(G, \ )$; φ_A is multiplication by the integer m: $a \to ma$ for a in A^G. Then multiplication by m is (obviously) a cohomological extension of φ and is the only one (by the uniqueness of Theorem 1).

(2) Restriction.

If H is a subgroup of G, then restriction of the action from G to H, gives a functor $\rho : Mod_G \to Mod_H$. Let

$F_1 = V$, $C_1 = H^*(G, \ )$; $F_2 : A \mapsto (A\rho)^H$, $C_2 = H^*(H, \cdot\, \rho)$;
φ_A = inclusion: $A^G \mapsto (A\rho)^H$.

The unique cohomological extension,

$$\text{res} : \ H^*(G,A) \mapsto H^*(H, A\rho)$$

is called restriction.

(3) Inflation.

Let $1 \to N \to G \xrightarrow{\pi} Q \to 1$ be exact and $\rho : \text{Mod}_Q \to \text{Mod}_G$ be determined by π. Essentially as in (2) we obtain a natural homomorphism, called lifting :

$$H^*(Q,M) \mapsto H^*(G, M\rho).$$

Now $A \mapsto A^N$ is a functor $\text{Mod}_G \to \text{Mod}_Q$ and so, for each $q \geq 0$, we have a functor $A \mapsto H^q(Q,A^N)$. The inclusion $(A^N)\rho \to A$ now yields a homomorphism, called inflation:

$$\text{inf}_A^q : H^q(Q, A^N) \xrightarrow{\text{lifting}} H^q(G, A^N\rho) \to H^q(G,A).$$

Clearly, inf^q is a natural homomorphism:

$$H^q(Q, \ .^N \ ) \to H^q(G, \ ).$$

We write $\text{inf} = (\text{inf}^q)$.

Inflation is simply a sequence of natural homomorphisms: it does not make sense to ask whether it is a natural homomorphism of cohomological functors since $H^*(Q, \ .^N)$ is not (in general) a cohomological functor.

(4) Corestriction.

Let H be a subgroup of finite index in G. Take $F_1 : A \mapsto A^H$, where A is a G-module; $C_1 = H^*(H, \ )$; $F_2 = V: A \mapsto A^G$; $C_2 = H^*(G, \ )$. To see that C_1 is minimal we observe first that, for any abelian group C, we have an isomorphism of H-modules:

$$\mathrm{Hom}_{\mathbb{Z}}(\mathbb{Z}G, C) \cong \prod_{t \in T} \mathrm{Hom}_{\mathbb{Z}}((\mathbb{Z}H)t, C),$$

where T is a right transversal of H in G. Then we use the simple

LEMMA 1. If (B_λ) is a family of H-modules,

$$H^q(H, \Pi B_\lambda) \cong \Pi\, H^q(H, B_\lambda), \quad \text{all } q \geq 0.$$

Proof = exercise. (Use the usual method of pulling up via a coinduced module.)

Define $\tau_{G|H} : A^H \mapsto A^G$ by

$$a \to \sum_{t \in T} at \ ,$$

where T is any right transversal of H in G. Clearly $\tau_{G|H}$ is independent of the choice of T.

The unique extension of $\tau_{G|H}$ is called corestriction:

$$\mathrm{cor} : H^*(H,A) \mapsto H^*(G,A).$$

PROPOSITION 1. If $(G:H) = k$, then $\mathrm{res} \circ \mathrm{cor} = k$.

Proof. True in dimension 0, therefore true by the uniqueness of Theorem 1. (A product of natural homomorphisms of cohomological functors is a ditto.)

COROLLARY 1. If G is finite of order k, then $kH^q(G,A) = 0$ for all $q > 0$ and all A.

Proof. Take $H = 1$ in the Proposition and use that $H^q(1,A) = 0$ for $q > 0$.

COROLLARY 2. If the prime p does not divide $k = (G:H)$, then res is one-one on the p-primary part of each $H^q(G,A)$, $q \geq 0$. (Application: G finite, H a p-Sylow.)

PROPOSITION 2. If G is finite and A is a G-module, finitely generated over $\mathbb{Z}$, then $H^q(G,A)$ is finite for all $q > 0$.

Proof. By using a finitely generated resolution of $\mathbb{Z}$ we see that all $H^q(G,A)$ are finitely generated. Hence the result by Corollary 1 above.

If H is a subgroup of G and A is an H-module, put

$$A\uparrow_H^G = \mathrm{Hom}_{\mathbb{Z}H}(\mathbb{Z}G, A),$$

viewed as a right G-module via the left module structure on $\mathbb{Z}G$. The functor $\uparrow_H^G : \mathrm{Mod}_H \to \mathrm{Mod}_G$ is exact and hence $(H^q(G, \uparrow_H^G))$ is a cohomological functor. Moreover, it is minimal. This follows from

LEMMA 2. For any abelian group C we have a G-module isomorphism

$$\underline{C\!\uparrow_1^G \;=\; (C\!\uparrow_1^H)\!\uparrow_H^G}\ .$$

Proof. Map f in $C\!\uparrow_1^G$ to $f^{\rightarrow}$, defined thus: for x in G,

$$xf^{\rightarrow} \;:\; h \rightarrow (xh)f, \quad h \in H.$$

Conversely, if $g \in (C\!\uparrow_1^H)\!\uparrow_H^G$, define $g^{\leftarrow}$ by:

$$x\, g^{\leftarrow} \;=\; 1(xg).$$

Then $f^{\rightarrow\leftarrow} = f$, $g^{\leftarrow\rightarrow} = g$ and one checks that these are G-homomorphisms.

Now $(H^q(G, \;\uparrow_H^G))$ extends $A \mapsto (A\!\uparrow_H^G)^G$. Clearly

$$\varphi_A \;:\; (A\!\uparrow_H^G)^G \;\tilde{\rightarrow}\; A^H$$

by $f \rightarrow 1f$; and φ is a natural isomorphism of functors. Hence (by Theorem 1), φ and φ^{-1} extends uniquely to natural homomorphisms ψ, ψ^{-1}, where

$$\psi \;:\; (H^q(G, \;\uparrow_H^G)) \rightarrow (H^q(H, \;)).$$

Thus we have

THEOREM 2. (Shapiro Lemma.) If H is a subgroup of G, then $(H^q(G, \;\uparrow_H^G))$, $(H^q(H, \;))$ are naturally isomorphic cohomological functors.

6.4 The inflation-restriction sequence.

An exact sequence of groups $1 \to N \to G \to Q \to 1$ yields the natural homomorphisms

$$\mathrm{res}_A^q : H^q(G,A) \mapsto H^q(N,A) ,$$

$$\mathrm{inf}_A^q : H^q(Q,A^N) \mapsto H^q(G,A) ,$$

for all G-modules A and all $q \geq 0$.

PROPOSITION 3. (i) The following sequence is exact:

$$0 \to H^1(Q,A^N) \xrightarrow{\mathrm{inf}} H^1(G,A) \xrightarrow{\mathrm{res}} H^1(N,A) .$$

(ii) If $q > 1$ and $H^i(N,A) = 0$ for $i = 1,\dots,q-1$, then the following is exact:

$$0 \to H^q(Q,A^N) \xrightarrow{\mathrm{inf}} H^q(G,A) \xrightarrow{\mathrm{res}} H^q(N,A).$$

We shall need this result in a number of places. The proof is quite straightforward and very accessible in Serre, Corps Locaux, pages 125, 126.

6.5 The trace map for finite groups.

Let G be a finite group and write

$$\tau = \tau_{G|1} = \sum_{x\in G} x.$$

Then, for any G-module A, $A\tau \leq A^G$.

Definition. A is called relatively injective (($\mathbb{Z}G$, $\mathbb{Z}$)-injective) if

$$0 \to A \xrightarrow{\mu} A^* \to A' \to 0$$

splits, where μ is the usual embedding: $a\mu$: $x \to ax$.

LEMMA 3. A is relatively injective if, and only if, there exists a $\mathbb{Z}$-endomorphism ρ of A such that $1_A = \rho^\tau$.

Proof. Assume $1_A = \rho^\tau$ and let ν: $A^* \to A$ be

$$f\nu = \sum_{x\in G} (x^{-1}f)\rho x.$$

Then ν is a G-homomorphism and $\mu\nu = 1_A$.

Conversely, let ν: $A^* \to A$ be a G-homomorphism such that $\mu\nu = 1_A$. Define $\varkappa : A \to A^*$ by

$$x(a\varkappa) = (\delta_{x,1})a.$$

So $\varkappa$ is a $\mathbb{Z}$-homomorphism and if $\rho = \varkappa\nu$, then $\rho^\tau = 1_A$.

LEMMA 4. Every coinduced module is relatively injective.

Proof. Let $\nu: C^{**} \to C^*$ be $x(\varphi\nu) = 1(x\varphi)$. Then $\mu\nu = 1$ (where $\mu : C^* \to C^{**}$) and ν is a G-homomorphism.

LEMMA 5. If A is relatively injective, then $A\tau = A^G$.

Proof. By Lemma 3, $1_A = \rho^\tau$ for some ρ. Hence, if $a \in A^G$,

$$a = a^{\rho^\tau} = \sum_x ax^{-1}\rho x = (a\rho)\tau \in A\tau .$$

PROPOSITION 4. For any G-module A, $H^1(G,A) \cong B^G/B\tau$, where $B = A^*/A\mu$ (μ being the usual embedding $A \to A^*$).

Proof. By Lemma 4, A^* is relatively injective and so, by Lemma 5, $(A^*)^G = (A^*)\tau$. But

$$H^1(G,A) = \mathrm{Coker}\ ((A^*)^G \to B^G)$$

and so the result.

CHAPTER 7

FINITE p-GROUPS

7.1 Frattini groups.

Throughout this section let G be a group admitting a set of operators P. This means that we are given a mapping of P into the set of endomorphisms of G.

Special case: P is a group and G is a P-group in the sense of Chapter 1 (§1.1).

Definition. The intersection with G of all maximal P-subgroups of G is called the P-Frattini group of G and will be written $Fr_P(G)$.

Examples.

(1) P is the empty set. Here we simply write Fr(G), called the Frattini group of G.

(2) P is a group, $G \triangleleft P$ and P acts by conjugation. Here $Fr_P(G)$ is the intersection of all K so that G/K is a chief factor of P.

(3) P is a ring, G is the additive group of P and P acts by right multiplication. Here P-subgroup = right ideal. Thus $Fr_P(G)$ is the Jacobson radical of P.

(4) P is a ring, G is a P-module. $Fr_P(G)$ is the intersection of all submodules H so that G/H is a simple module. Hence we have $Fr_P(G) \geq GJ$, where J is the Jacobson radical of P.

<u>Definition</u>. We call x in G a <u>non-generator</u> if $G = \langle P \langle S, x \rangle$ (= P-admissible subgroup generated by S and x) always implies $G = \langle P \langle S \rangle$.

The set of all non-generators is obviously a P-admissible subgroup that admits all P-automorphisms of G.

<u>PROPOSITION 1. The set of all non-generators coincides with $Fr_P(G)$.</u>

<u>Proof</u> = exercise.

(Note that the "Nakayama Lemma" is a very special case of this.)

If G is finitely P-generated (i.e., finitely generated as a group admitting P), we define $d_P(G)$ to be the <u>minimum number of P-generators</u> of G. When $P = \phi$ write $d(G)$ for this number.

An immediate consequence of Proposition 1 is the

<u>COROLLARY. If G and $Fr_P(G)$ are finitely P-generated then</u>

$$G = \langle P \langle S \rangle \, Fr_P(G)$$

<u>implies</u> $$G = \langle P \langle S \rangle \; ;$$

<u>and so</u> $$d_P(G) = d_P(G/Fr_P(G)) \; .$$

<u>Exercise</u>. If $H \triangleleft G$ and H is abelian, then $Fr(G) \geq Fr_G(H)$. (Hint: For each maximal subgroup M of G, $H/H \cap M$ is a chief factor of G.) The inequality is a special case of a much more general result: cf. Lemma 3 in [4].

Definition. If $H \lhd G$ and both admit P, we say H has a P-supplement S (is P-supplemented in G) if S is a P-subgroup so that G = SH and $S \neq G$.

PROPOSITION 2. If G is such that every P-subgroup is contained in a maximal P-subgroup, then H is supplemented if, and only if, $H \not\leq Fr_P(G)$.

Proof. = exercise.

7.2 Generators and relations for p-groups.

Throughout this section G will be a finite p-group.

Since every maximal subgroup of G is normal, $Fr(G) = G^*$, where $G^* = G^pG'$. Hence, if $b_1G^*, \ldots, b_nG^*$ is a basis of the vector space G/G^*, $b_1, \ldots, b_n$ alone generate G and $d(G) = \dim G/G^*$ (Corollary to Proposition 1).

Using §3.5 we have

PROPOSITION 3.

$$H^1(G, \mathbb{F}_p) = \mathrm{Hom}(G, \mathbb{F}_p) ;$$

$$H_1(G, \mathbb{F}_p) = G/G^* ;$$

and the dimension of each as a vector space over $\mathbb{F}_p$ is d(G).

Choose a free group F on n = d(G) generators. Map the free generators to any elements in G that form a basis of G mod G^*.

This extends to a homomorphism which is necessarily surjective. So we have a presentation

$$1 \to R \to F \to G \to 1,$$

where $R \leq F^* = F^pF'$. We call such a presentation a <u>minimal presentation</u>.

By Mac Lane's result (Proposition 6, §3.6), $H^2(G, \mathbb{F}_p)$ is the cokernel of

$$\mathrm{Der}(F, \mathbb{F}_p) \to \mathrm{Hom}_G(R/R', \mathbb{F}_p).$$

Clearly, $\mathrm{Hom}_G(R/R', \mathbb{F}_p) = \mathrm{Hom}(R/R^\#, \mathbb{F}_p)$, where $R^\# = R^p[R,F]$; and $\mathrm{Der}(F, \mathbb{F}_p) = \mathrm{Hom}(F/F^*, \mathbb{F}_p)$. Since $R \leq F^*$, we have the first formula of the next proposition. The second formula is a consequence of Proposition 7, §3.6.

<u>PROPOSITION 4</u>.
$$\underline{H^2(G, \mathbb{F}_p) = \mathrm{Hom}(R/R^\#, \mathbb{F}_p)\ ;}$$
$$\underline{H_2(G, \mathbb{F}_p) = R/R^\#\ .}$$

It follows that $d(R/R^\#) = \dim R/R^\#$ is a constant for all minimal presentations. We shall write this number as $r(G)$. On the face of it, one could imagine that a non-minimal presentation might give a smaller $\dim R/R^\#$. That this cannot, in fact, happen is implied by

<u>PROPOSITION 5. If $1 \to S \to E \to G \to 1$ is any presentation of G over a free group E, then $d(E) \geq d(G)$ and $\dim S/S^\# \geq r(G)$.</u>

<u>Proof</u>. Since E is free, there exists $\theta : E \to F$ so that

$$\begin{array}{ccc} E & \xrightarrow{\theta} & F \\ & \searrow \swarrow & \\ & G & \end{array}$$

commutes. Then $E^\theta R = F$ implies $E^\theta F^* \equiv F \pmod{R^\#}$. Since $F/R^\#$ is a finite p-group with Frattini group $F^*/R^\#$, we conclude $E^\theta \equiv F \pmod{R^\#}$. Thus θ induces a surjection : $E/E^* \to F/F^*$; and also a surjection $S/S^\# \to R/R^\#$.

Clearly

$$S/[S,E] \cong (S/S\cap E') \times (S\cap E'/[S,E]),$$

since the first factor is free abelian. The second factor is $H_2(G,\mathbb{Z})$ and so is a finite p-group. This follows from Proposition 2, §6.2 together with the universal coefficient theorem (Theorem 3, §3.7). Alternatively one can see it directly by taking the transfer of $F/[R,F]$ into $R/[R,F]$ (cf. Schenkman, p. 134 or Scott, p. 60). Thus

$$d(S/[S,E]) = d(S/S^\#) = d(E) + d(H_2(G, \mathbb{Z})).$$

PROPOSITION 6. *$d(S/S^\#) - d(E)$ is a constant for all free presentations of G and is equal to $d(H_2(G,\mathbb{Z}))$.*

The Proposition also implies that $d(S/S^\#) = r(G)$ if, and only if, the presentation is minimal.

The integral duality theorem (Cartan-Eilenberg, p. 250), implies

$$d(H_2(G,\mathbb{Z})) = d(H^3(G,\mathbb{Z}))$$

and so we have yet another interpretation of $r(G) - d(G)$. It is easy to give a direct proof:

$$0 \to \mathbb{Z} \xrightarrow{p} \mathbb{Z} \to \mathbb{F}_p \to 0$$

gives the exact sequence:

$$0 \to H^1(G,\mathbb{F}_p) \to H^2(G,\mathbb{Z}) \xrightarrow{p} H^2(G,\mathbb{Z}) \to H^2(G,\mathbb{F}_p) \to H^3(G,\mathbb{Z})_p \to 0$$

since $H^1(G,\mathbb{Z}) = \mathrm{Hom}(G,\mathbb{Z}) = 0$ (G finite!) and where $H^3(G,\mathbb{Z})_p$ denotes the subgroup of elements of order p. All groups are finite (Proposition 2, §6.2) and so take the alternating product of the orders.

The invariant r(G) has a good interpretation in the category of pro-p-groups. Consider again a minimal presentation

$$1 \to R \to F \to G \to 1.$$

Let $\hat{F} = \varprojlim F/P$, where P runs through all normal subgroups of p-power index in F. Then $\hat{F}$ is a pro-p-group. Note that, by Theorem 2 of §4.3, $F \to \hat{F}$ is injective. Defining $\hat{R}$ similarly, it is easy to verify that

$$1 \to \hat{R} \to \hat{F} \to G \to 1$$

is an exact sequence of pro-p-groups (so that, in particular, the maps are continuous homorphisms).

PROPOSITION 7. r(G) is the minimum number of relations needed to define G as a pro-p-group.

LEMMA 1. *Let $\mathcal{P}$ be the set of all normal P in F such that $P \leq R$ and F/P is a finite p-group. Choose a basis B of R modulo $R^{\#}$. Then $R = \langle F\langle B\rangle P$ (= normal closure of B and P) for all P in $\mathcal{P}$; and*

$$r(G) = \max\{d_F(R/P) \mid P \in \mathcal{P}\}.$$

Proof. If $P \in \mathcal{P}$, $Fr_F(R/P) = R/R^{\#}P$ and $d_F(R/R^{\#}P) = d(R/R^{\#}P)$. Hence the result follows by the Corollary to Proposition 1.

Proof of Proposition 7. Let t be the minimum number of relations and B any basis of R modulo $R^{\#}$. By Lemma 1, $\langle\hat{F}\langle B\rangle$ is dense in $\hat{R}$. Hence $t \leq \text{card } B = r(G)$.

Conversely, $t \geq d_F(R/P)$ for all P in $\mathcal{P}$ ($\mathcal{P}$ as in Lemma 1) and so $t \geq r(G)$.

Lemma 1 implies that $r(G) = d_G(R/R^*)$ (where $R^* = R^pR'$). As a matter of fact, if $1 \to S \to E \to G \to 1$ is a second minimal presentation, it is easy to prove that $S/S^* \cong R/R^*$ as $\mathbb{F}_pG$-modules. We prefer to postpone the proof until the chapter on extension categories, where it will appear in a more general setting.

The minimum number of relations needed to define G as an abstract p-group is $d_F(R)$. Clearly $d_F(R) \geq r(G)$.

Problems: (1) Is $d_F(R)$ an invariant for G?

(2) What is the class of p-groups G for which $d_F(R) = r(G)$?

7.3 The Golod-Šafarevič inequality.

Šafarevič [7] showed that a famous problem concerning Hilbert class fields (cf. §7.4, below) can be reduced to a problem about the dimension and "p-relation number" of finite p-groups. The problem that had to be decided was this:

$$\text{Does} \quad r(G) - d(G) \longrightarrow \infty \quad \text{as} \quad d(G) \to \infty \ ?$$

Answer: Yes. This was first established by Golod-Šafarevič [3]. The following inequality is a slight improvement (by now well-known) of their result.

<u>THEOREM 1. For any finite p-group G,</u>

$$\underline{r(G) > \frac{d(G)^2}{4}} \ .$$

The connexion between this result and the class field tower problem is not a trivial matter and we cannot usefully say anything about it in this course. We refer to Serre's notes [8] or the article by Roquette [6].

Theorem 1, on the other hand, is essentially an elementary fact.

<u>Proof</u>. (<u>Roquette</u>) Take a minimal presentation $1 \to R \to F \xrightarrow{\pi} G \to 1$ of G and consider

$$0 \to \mathfrak{r}/\mathfrak{f}\mathfrak{r} \to \mathfrak{f}/\mathfrak{f}\mathfrak{r} \to \mathfrak{g} \to 0$$

taken over $\mathbb{F}_p$. The minimality implies

$$\mathfrak{r} \leq \mathfrak{f}^2 . \tag{1}$$

Write $B = \mathfrak{f}/\mathfrak{f}\mathfrak{r}$ and A for the free $\mathbb{F}_p G$-module on some set $a_1,\ldots,a_r$ $(r = r(G))$. Let $\alpha\colon\ A \to \mathfrak{r}/\mathfrak{f}\mathfrak{r}$ be the epimorphism

$$a_i \mapsto (1-b_i) + \mathfrak{f}\mathfrak{r} ,$$

where $b_1,\ldots,b_r$ form a basis of R modulo R^*. Then

$$A \xrightarrow{\alpha} B \xrightarrow{\pi} \mathfrak{g} \to 0 \tag{2}$$

is exact.

If we put

$$A_k = \{ a\in A \mid a\alpha\in B\mathfrak{g}^k\} ,$$

then

$$0 \to A_k/A_{k+1} \to B\mathfrak{g}^k/B\mathfrak{g}^{k+1} \to \mathfrak{g}^{k+1}/\mathfrak{g}^{k+2} \to 0 \tag{3}$$

is exact. Let

$$d_k = \dim \mathfrak{g}^k/\mathfrak{g}^{k+1} ,$$

$$e_k = \dim A_k/A_{k+1} .$$

Then $d_0 = 1$, $d_1 = d = d(G)$ and $e_0 = 0$ because $A\alpha \leq B\mathfrak{g}$ (by (1)). Moreover, $\dim B\mathfrak{g}^k/B\mathfrak{g}^{k+1} = dd_k$, because B is free of rank d. Now (3) gives

$$dd_k = d_{k+1} + e_k. \tag{4}$$

Since $\mathbb{F}_p G$ is finite dimensional, $d_k = 0$ for all sufficiently large k. (Actually $\mathfrak{g}$ is even nilpotent but we do not need this fact.) If we put

$$f(X) = \sum_{k\geq 0} d_k X^k ,$$

$$g(X) = \sum_{k\geq 0} e_k X^k ,$$

then both f and g are really polynomials.

By (1), $A\mathfrak{g}^k \leq A_{k+1}$ for all $k \geq 0$. Hence

$$e_1 + \dots + e_k \leq r(d_0 + \dots + d_{k-1})$$

and so

$$g(t) / (1-t) \leq r\, t\, f(t) / (1-t)$$

for all $0 < t < 1$. Thus

$$g(t) \leq r\, t\, f(t). \tag{5}$$

Multiply (4) by X^{k+1} and add:

$$d\, X\, f(X) = f(X) - 1 + X\, g(X).$$

Hence, by (5),

$$1 \leq f(t)\,(1 - dt + rt^2)$$

for all $0 < t < 1$. Since all the coefficients of $f(X)$ are non-negative,

$$0 < 1 - dt + rt^2 ,$$

and so the discriminant of this must be negative:

$$d^2 - 4r < 0,$$

as required.

Gaschütz has proved an interesting generalization of Theorem 1 valid for any finite supersoluble group G.

Ingredients in the Gaschütz formula:

(1) The irreducible $\mathbb{F}_pG$-modules in the block containing $\mathbb{F}_p$ form a group with respect to $\otimes$: let h be the order of this group. (If G is a p-group, h = 1.)

(2) Let P be the indecomposable projective $\mathbb{F}_pG$-module so that $P/Fr_G(P) \cong \mathbb{F}_p$ and let Q be the kernel of any minimal epimorphism: $P_1 \to Fr_G(P)$. Write $r = d_G(Q)$. (If G is a p-group, $r = r(G)$.)

(3) Let c be the number of complemented p-chief factors of G. (If G is a p-group, $c = d(G)$.)

Then the Gaschütz formula is

$$r > \frac{c^2}{4h} .$$

7.4 Hilbert class fields.

Let R be an integral domain and K its quotient field. A non-zero R-submodule M of K is called a fractional ideal (frid) of R if there exists $x \neq 0$ in R such that $xM \leq R$.

Let Id_K be the set of all such frids. This is a semi-group. If Id_K is a group, R is called a Dedekind domain. We assume this henceforth. (The identity of Id_K must be R.)

For each $x \in K^*$ (= non-zero elements in K), xR is a frid and these form a subgroup, the group of principal ideals. Note $xR = R$ if, and only if, x is an invertible element of R. So we have the exact sequence of abelian groups:

$$1 \to U(R) \to K^* \to Id_K \to Cl_K \to 1.$$

($U(R)$ is the group of invertible elements in R.) The group Cl_K is called the ideal class group of R.

Note: $Id_K = 1$ if, and only if, R is a field and $Cl_K = 1$ if, and only if, R is a principal ideal domain. Cl_K measures how far R is from being a principal ideal domain.

Now let F be an algebraic number field (= finite extension of $\mathbb{Q}$) and A the ring of algebraic integers in F. Then A is a Dedekind domain (e.g. [10], chapter 5). A basic theorem of arithmetic asserts that Cl_F is finite.

Let E be a finite extension of F and B its ring of algebraic integers. Then $F \to E$ induces natural homomorphisms $Id_F \to Id_E$ (by $\mathfrak{a} \to \mathfrak{a}B$) and $v_{E/F} : Cl_F \to Cl_E$.

Hilbert conjectured and Furtwängler proved that to each F there exists a finite Galois extension E/F such that $Gal(E/F) \cong Cl_F$ and $v_{E/F}$ is the trivial homomorphism (i.e., all frids of F become principal in E). We call E the Hilbert class field of F.

Let L be the Hilbert class field of E, where E is the Hilbert class field of F. Thus $G = Gal(L/F)$ is metabelian and $Gal(L/E)$ must

$$
\begin{array}{lll}
L \bullet & \bullet\, 1 & \\
E \bullet & \bullet\ \mathrm{Gal}(L/E) \cong \mathrm{Cl}_E & \bullet\ 1 \\
F \bullet & \bullet\ \mathrm{Gal}(L/F) = G & \bullet\ \mathrm{Gal}(E/F) \cong \mathrm{Cl}_F
\end{array}
$$

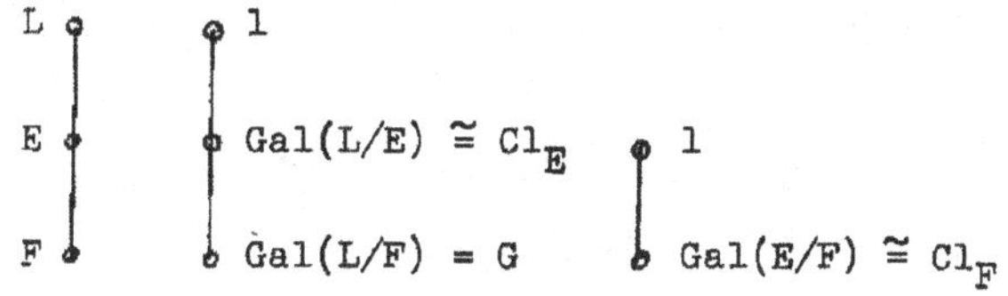

be $G' = [G,G]$ (because one can show that E/F is a <u>maximal</u> abelian extension). Now the arithmetically defined homomorphism $v_{E/F}$ yields a group theoretical homomorphism: $G/G' \to G'$. This is precisely the group theoretical transfer and the triviality of this group homomorphism (e.g., [11], Chapter 5, §4) establishes the last part of Hilbert's conjecture.

Given an algebraic number field F, denote its Hilbert class field by Hil(F). The old problem concerning these was this: given F, does there always exist k such that $\mathrm{Hil}^k(F) = \mathrm{Hil}^{k+1}(F)$? The Šafarevič reduction together with Theorem 1 above give a negative answer:

The ŠAFAREVIČ THEOREM. There exists an algebraic number field F having an infinite class field tower (i.e., $\mathrm{Hil}^{k+1}(F) > \mathrm{Hil}^k(F)$ for $k = 0, 1, 2, 3, \dots$).

An explicit example of such a field F is $\mathbb{Q}\sqrt{-3.5.7.11.13.17.19}$.

7.5 Outer automorphisms of order p.

One of the most ingenious applications of cohomology to a purely group theoretical problem is the recent solution by Gaschütz of the question whether every finite p-group has outer automorphisms of order p.

THEOREM 2 (Gaschütz [1], [2].) If G is a non-simple finite p-group then p divides $|\mathrm{Out}\ G|$.

Example. If C is a cyclic group of order p and $G = C[(\mathbf{F}_p C)$, then $|\mathrm{Out}\ G| = p$. Note that G is (isomorphic to) a p-Sylow subgroup of the symmetric group of degree p^2 : $G \cong C \wr C$ (wreath product).

Theorem 2 is obviously true for non-simple abelian groups. We henceforth assume G is non-abelian.

The following key lemma is needed and will be proved below after we are done with the main argument.

LEMMA 2. (Gaschütz [1], Uchida [9].) Let H be a finite p-group and A an H-module that is also a finite p-group. If $H^1(H,A) = 0$, then $H^k(S,A) = 0$ for all $k \geq 1$ and all $S \leq H$.

We also need the very simple

<u>LEMMA 3. If A is a maximal abelian normal subgroup of a nilpotent group G, then $A = C_G(A)$ (centralizer).</u>

<u>Proof</u>. Suppose $C = C_G(A) > A$. Define $C_0 = C$ and inductively $C_{i+1} = [C_i, G]$ for $i \geq 0$. Let k be the first integer such that $C_{k+1} \leq A$. Choose $x \in C_k$, $x \notin A$. Then $B = \langle x, A \rangle$ is an abelian normal subgroup strictly bigger than A: a contradiction.

<u>Proof of Theorem 2</u>.

<u>First Step</u>. Let A be a maximal abelian normal subgroup of G. If D is all automorphisms of G fixing A and G/A pointwise and D_0 is the image of A in Aut G, then

$$D_0 = D \cap \mathrm{In}\, G$$

by Lemma 3. Recall that

$$D/D_0 \cong H^1(G/A, A)$$

(Proposition 5, §3.5) and this is a finite p-group. Therefore we are done if $D \neq D_0$. Henceforth assume

$$(*) \qquad H^1(G/A, A) = 0 \quad \text{for all maximal abelian normal } A.$$

<u>Second Step</u>. Suppose there exists a maximal subgroup N so that $\zeta_1(N) \leq \zeta_1(G)$. Then N is normal, G/N is of order p and $\zeta_1(N)$ contains a subgroup of order p. We can therefore pick a homomorphism f of G into $\zeta_1(N)$ so that $N = \mathrm{Ker}\, f$. Put

$$x^{\varphi} = x\,x^{f}\,.$$

Then φ is an endomorphism (because x^f is central) and one-one (because $x^{f^2} = 1$): i.e., $\varphi \in \operatorname{Aut} G$. Moreover, φ has order p (because $(x^f)^p = 1$).

If φ were conjugation by g, then $g \in C = C_G(N)$ (because φ fixes all elements of N). If $C \nleq N$, $G = CN$ and C is abelian (because $C/\zeta_1(N) \cong CN/N$ is cyclic). Hence $g \in \zeta_1(G)$. But if $C \leq N$, $C = \zeta_1(N) \leq \zeta_1(G)$ (by hypothesis) and again $g \in \zeta_1(G)$. In any case, conjugation by g is the identity. Consequently φ is really an outer automorphism.

Henceforth we assume:

(**) For every maximal subgroup N, $\zeta_1(N) \nleq \zeta_1(G)$.

<u>Third Step</u>. Take any maximal abelian normal subgroup A. By (*) and Lemma 2, $H^2(G/A,A) = 0$ and so (extension theory) A has a complement L: $G = L[A$. Pick a maximal subgroup N containing L. By (**), $\zeta_1(N) \nleq B = A \cap N$. Hence there exists $R \leq \zeta_1(N)$ so that $(R:C) = p$ (where $C = B \cap \zeta_1(N)$). Set $S = AR/A$. By Lemma 2, $H^2(S,A) = 0$ and so $A^S = A\tau$ (§3.3), where $\tau = \tau_S$ (trace for S). Since $A^S \neq A$ (Lemma 3) and $B \leq A^S$, $A^S = B$. If $A = \langle u,B\rangle$, then

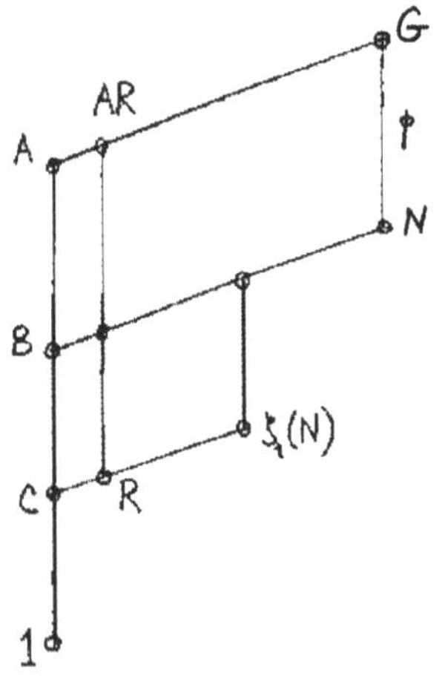

$$B = A\tau = \langle u\tau,\ B\tau\rangle = \langle u\tau,\ B^p\rangle = \langle u\tau\rangle ,$$

because $B^p \leq Fr(B)$.

Fourth Step. Suppose all abelian normal subgroups of G are cyclic. Then A and $\zeta_1(N) = Z$ in the third step are cyclic. Since $G = L[A$, $N = L[B$ and thus $B \neq C$ (otherwise $Z = (Z \cap L) \times B$).

Choose $x \in B$, $x \notin C$ so that $[x,N] \leq C$ (cf. proof of Lemma 3). Thus $M = \langle x, C\rangle$ is normal in G, whence $MZ \lhd G$. But MZ is abelian and therefore cyclic (hypothesis). Yet $MZ/C \cong M/C \times Z/C$ and both factors on the right hand side are non-trivial.

Hence not all abelian normal subgroups of G can be cyclic and we may assume our A in the third step is non-cyclic.

Fifth Step. Returning to Step 3 and knowing A is not cyclic, we have $A = B \times \langle a\rangle$, where $a^p = 1$. Let s generate $S = RA/A$. Then $a^s = ay$, for some y in B. As $s^p = 1$, $y^p = 1$ and

$$a\tau = a^p\, y^{1+2+\ldots+(p-1)} = y^{\frac{1}{2}p(p-1)} .$$

Since $B = \langle a\tau\rangle$ (cf. end of third step), we must have $p \nmid \frac{1}{2}p(p-1)$. Hence $p = 2$ and $|G| = 8$. Thus G is the quaternion group or the dihedral group. For these groups the theorem is true.

Proof of Lemma 2.

First Step. (Hoechsmann [5].) We assert that the hypothesis implies $H^1(N,A) = 0$ for all maximal subgroups N.

<u>Proof</u>. Embed A in the H-coinduced A^* ($= A\uparrow_1^H$): say

$$0 \to A \to A^* \to B \to 0.$$

By Proposition 4, §6.5,

$$B^H = B\tau_H \tag{1}$$

(where $\tau_H = \tau_{H|1}$, the trace map for H). We assert

$$B^N = B\tau_N . \tag{2}$$

This will prove what we want since A^* is a product of (H:N) copies of $A\uparrow_1^N$ (the N-coinduced module determined by A): cf. §6.5 and Lemma 1, §6.2.

Let C = H/N be generated by c and take any b in B^N. Then $b\tau_C \in (B^N)^C = B^H$ and so, by (1), $b\tau_C = b'\tau_H$ for some b' in B. Thus

$$(b - b'\tau_N)\tau_C = 0 \tag{3}$$

and therefore $b - b'\tau_N \in (B^N)_{\tau_C}$.

But $(B^N)^C = B\tau_H$ (by (1)) and

$$B\tau_H = (B\tau_N)\tau_C \leq B^N\tau_C .$$

So $(B^N)^C = B^N\tau_C$, whence $(B^N)_{\tau_C} = (B^N)(1-c)$, by Lemma 6, §3.3. Now (3) implies

$$b-b'\tau_N = b''(1-c),$$

for some b'' in B^N. Thus

$$b \equiv b''(1-c) \quad (\text{mod } B\tau_N).$$

Now $c^p = 1$ and hence $B^N = pB^N + B\tau_N$. Since $pB^N = Fr(B^N)$, we have (2) as required.

<u>Second Step</u>. We prove the lemma by an induction on $|H|$. By the first step and the induction, $H^k(S,A) = 0$ for all $k \geq 1$ and all $S < H$.

It remains to see $H^k(H,A) = 0$ for all $k > 1$. Take any maximal N. Then $H^1(H/N, A^N) = 0$ by Proposition 3 (i), §6.4, and consequently $H^k(H/N, A^N) = 0$ for all $k \geq 1$, by Lemma 6, §3.3. We also know $H^k(N,A) = 0$ for all $k \geq 1$. Hence, by Proposition 3 (ii), §6.4, $H^k(H,A) = 0$ for all $k \geq 1$.

Sources and References.

§7.2 is based largely on Serre's treatment of pro-p-groups in [8], chapter I, §4. The proof of Theorem 1 (§7.3) is taken from a set of notes by Roquette, circulated in the spring of 1964. The remarks on the Gaschütz generalization (end of §7.3) are based on a lecture he gave in London, spring 1965.

[1] Gaschütz, W.: Kohomologische Trivialitäten und äussere Automorphismen von p-Gruppen, Math. Zeit, 88 (1965) 432-433.

[2] Gaschütz, W.: Nichtabelsche p-Gruppen besitzen äussere p-Automorphismen, J. of Algebra 4 (1966) 1-2.

[3] Golod, E.S. and Šafarevič, I.: On the class field tower, Izv. Akad. Nauk SSSR 28 (1964) 261-272.

[4] Hall, P.: The Frattini subgroup of finitely generated groups, Proc. London Math. Soc. 11 (1961) 327-352.

[5] Hoechsmann, K.: An elementary proof of a lemma by Gaschütz, Math. Zeit. 96 (1967) 214-215.

[6] Roquette, P.: Proc. of conference on algebraic number theory, Brighton, 1965, Academic Press, 1967.

[7] Šafarevič, I.: Algebraic number fields, Proc. of Stockholm Congress, 1962, 163-176.

[8] Serre, J.-P.: Cohomologie Galoisienne, Springer notes, 1965.

[9] Uchida, K.: On Tannaka's conjecture on the cohomologically trivial modules, Proc. Japan Acad. 41 (1965) 249-253.

[10] Zariski, O. and Samuel, P.: Commutative Algebra, volume I, Van Nostrand, 1958.

[11] Zassenhaus, H.: The theory of groups, Chelsea 1949.

CHAPTER 8

COHOMOLOGICAL DIMENSION

8.1 Definition and elementary facts.

Definition. A group G has cohomological dimension k (we write cd G = k) if $H^q(G,A) = 0$ for all $q > k$ and all A, but there exists a G-module A such that $H^k(G,A) \neq 0$.

PROPOSITION 1. The following are equivalent:

(i) cd $G \leq k$;

(ii) $H^{k+1}(G,A) = 0$ for all A;

(iii) if $\ldots \to P_2 \to P_1 \to P_0 \to \mathbb{Z} \to 0$ is any projective resolution of $\mathbb{Z}$ and $Y = \operatorname{Im}(P_k \to P_{k-1})$, then Y is projective. (When k = 0, interpret P_{-1} to be $\mathbb{Z}$.)

(iv) There exists a projective resolution of $\mathbb{Z}$ of length k: $0 \to P_k \to \ldots \to P_0 \to \mathbb{Z} \to 0$.

Proof. All is obvious except that (ii) implies (iii):

If k = 0 and we are given an exact sequence

$$0 \to A \to B \to C \to 0, \tag{1}$$

then the cohomology sequence and (ii) give that

$$B^G \to C^G$$

is surjective, i.e., that

$$\operatorname{Hom}_G(\mathbb{Z},B) \to \operatorname{Hom}_G(\mathbb{Z},C)$$

is surjective, i.e., that $\mathbb{Z}$ is G-projective.

Let $k > 0$ and take any short exact sequence (1). Then by Proposition 1, §2.1 (p.20),

$$\begin{array}{ccccccc} \mathrm{Hom}_G(P_{k-1},B) & \to & \mathrm{Hom}_G(Y,B) & \to & H^k(G,B) & \to & 0 \\ \downarrow & & \downarrow & & \downarrow & & \\ \mathrm{Hom}_G(P_{k-1},C) & \to & \mathrm{Hom}_G(Y,C) & \to & H^k(G,C) & \to & 0 \end{array}$$

has exact rows. The left hand down map is surjective as P_{k-1} is projective, the right hand down map is surjective by (ii) (it is part of the exact cohomology sequence from (1)) and so the middle map is surjective. Therefore Y is projective.

Exercise. cd $G = 0$ if, and only if, $G = 1$.

By a remark made earlier (Remark (1), p.35), all free groups have cohomological dimension $\leqslant 1$. Whether the converse is true had been an open problem for a long time. But in 1968 John Stallings [27] managed to prove that all groups of cohomological dimension $\leqslant 1$ are locally free and shortly thereafter Swan [32] was able to use Stalling's result to settle the question completely: All groups of cohomological dimension $\leqslant 1$ are free.

For every positive integer k there do exist groups of cohomological dimension k: any free abelian group of rank k has cohomological dimension precisely k (see §8.8 below).

The general group-theoretic significance of finite

cohomological dimension is still an almost untouched problem. The following is a useful (but trivial) general observation:

<u>PROPOSITION 2. (i) If cd G $\leqslant$ k, then cd H $\leqslant$ k for all subgroups H.</u> <u>(ii) All groups of finite cohomological dimension are torsion-free.</u>

<u>Proof</u>. (i) is immediate from Shapiro's Lemma (§6.3, p.92); (ii) follows from (i) and the fact that finite cyclic groups do not have finite cohomological dimension (§3.3, p.39).

The converse of (ii) is false: a free abelian group of infinite rank has infinite cohomological dimension (see §8.8 below).

<u>Exercises</u>.

1. Suppose H_1, H_2 are subgroups of G whose indices are finite and coprime. If cd $H_i \leqslant n$, i = 1, 2, then cd G $\leqslant$ n. (Hint: Use res o cor: p.90.)

<u>Definition</u>. If $H_q(G,A) = 0$ for all A, we say G has <u>homological dimension</u> (or <u>weak dimension</u>) $\leqslant$ k: hd G $\leqslant$ k.

2. Prove that the following are equivalent:

(i) hd G $\leqslant$ k;

(ii) $H_{k+1}(G,A) = 0$ for all A;

(iii) if $\ldots \to P_1 \to P_o \to \mathbb{Z} \to 0$ is a projective resolution of $\mathbb{Z}$ and $Y = \mathrm{Im}(P_k \to P_{k-1})$, then Y is <u>$\mathbb{Z}G$-flat</u> (i.e., $0 \to A \to B$ exact always implies $0 \to Y \underset{G}{\otimes} A \to Y \underset{G}{\otimes} B$ is exact).

(Note: (iii) implies that there exists a <u>flat resolution</u> of $\mathbb{Z}$ of length k. Then (iii) $\Rightarrow$ (i) comes from dimension shifting and the appropriate exact Tor sequences.)

3. (i) Prove that hd $G \leqslant$ cd G.

(ii) If G is a free group, hd G = cd G.

(iii) If G is the additive group of rationals, hd G = cd $G - 1$. (Use Proposition 4 in §8.3, below.)

<u>Problem</u>: Is cd $G - 1 \leqslant$ hd G always?

8.2 Test elements.

There exist "test elements" for finite cohomological dimension in a sense that we now wish to make precise (Proposition 3, below).

Take a free presentation $1 \to R \to F \to G \to 1$ and the corresponding resolution (Theorem 2, p.34). Recall that

$$H^{2n}(G, \mathfrak{r}^n/\mathfrak{f}\mathfrak{r}^n) =$$

$$\mathrm{Coker}(\mathrm{Hom}_G(\mathfrak{f}\mathfrak{r}^{n-1}/\mathfrak{f}\mathfrak{r}^n, \ \mathfrak{r}^n/\mathfrak{f}\mathfrak{r}^n) \to \mathrm{Hom}_G(\mathfrak{r}^n/\mathfrak{f}\mathfrak{r}^n, \ \mathfrak{r}^n/\mathfrak{f}\mathfrak{r}^n)).$$

Let the identity on $\mathfrak{r}^n/\mathfrak{f}\mathfrak{r}^n$ induce the cohomology class χ_{2n}.

If cd $G \leqslant 2n-1$, then of course $\chi_{2n} = 0$. Conversely, suppose $\chi_{2n} = 0$. Then the identity map on $\mathfrak{r}^n/\mathfrak{f}\mathfrak{r}^n$ is induced by some endomorphism φ of $\mathfrak{f}\mathfrak{r}^{n-1}/\mathfrak{f}\mathfrak{r}^n$ with image $\mathfrak{r}^n/\mathfrak{f}\mathfrak{r}^n$: so $\mathfrak{r}^n/\mathfrak{f}\mathfrak{r}^n$ is a direct summand of $\mathfrak{f}\mathfrak{r}^{n-1}/\mathfrak{f}\mathfrak{r}^n$, therefore $\mathfrak{f}\mathfrak{r}^{n-1}/\mathfrak{r}^n$ is isomorphic to a direct summand and is thus projective. Hence, by Proposition 1, (iii), cd $G \leqslant 2n-1$.

Next let the identity on $\mathfrak{f}\mathfrak{r}^n/\mathfrak{r}^{n+1}$ induce the cohomology class χ_{2n+1} in $H^{2n+1}(G, \mathfrak{f}\mathfrak{r}^n/\mathfrak{r}^{n+1})$. As above we see that cd $G \leqslant 2n$ if, and only if, $\chi_{2n+1} = 0$.

If χ_1 is the element of $H^1(G, \mathfrak{g})$ induced by the identity on $\mathfrak{g}$, then clearly cd $G = 0$ if and only if $\chi_1 = 0$.

Thus we have proved

PROPOSITION 3. cd $G \leqslant k$ if and only if $\chi_{k+1} = 0$ $(k \geqslant 0)$.

It is easy to determine the group theoretic significance of χ_2:

LEMMA 1. Let χ be the cohomology class corresponding to the extension $1 \to R/R' \to F/R' \to G \to 1$. Then the natural isomorphism $(1-r) + \mathfrak{f}\mathfrak{r} \mapsto rR'$ of $\mathfrak{r}/\mathfrak{f}\mathfrak{r} \xrightarrow{\sim} R/R'$ induces

$$H^2(G, \mathfrak{r}/\mathfrak{f}\mathfrak{r}) \xrightarrow{\sim} H^2(G, R/R')$$

and under this, $\chi_2 \mapsto \chi$.

Proof. The argument at the top of page 71 (§5.3) shows that χ is induced from the natural projection $F \to F/R'$ restricted to R: $r \mapsto rR'$, for r in R. Under

$$\mathrm{Hom}_F(R,\ R/R') \xrightarrow{\sim} \mathrm{Hom}_G(R/R',\ R/R'),$$

this projection corresponds to $1_{R/R'}$. Hence the result follows from the commutative square

$$\begin{array}{ccc}\mathrm{Hom}_G(\mathfrak{r}/\mathfrak{f}\mathfrak{r},\ \mathfrak{r}/\mathfrak{f}\mathfrak{r}) & \longrightarrow & \mathrm{Hom}_G(R/R',\ R/R') \\ \downarrow & & \downarrow \\ H^2(G,\ \mathfrak{r}/\mathfrak{f}\mathfrak{r}) & \longrightarrow & H^2(G,\ R/R')\ .\end{array}$$

One often calls χ the <u>characteristic class</u> of the presentation $1 \to R \to F \to G \to 1$.

We can obtain a more striking form of Proposition 3. For this purpose (and only this) we assume some knowledge of the cup product.

The G-module isomorphism

$$\mathfrak{r}^n/\mathfrak{f}\mathfrak{r}^n \xrightarrow{\sim} \underbrace{R/R' \otimes \ldots \otimes R/R'}_{n}$$

(Proposition 3, p.38) yields

$$H^{2n}(G,\ \mathfrak{r}^n/\mathfrak{f}\mathfrak{r}^n) \xrightarrow{\sim} H^{2n}(G,\ R/R' \otimes \ldots \otimes R/R')$$

and in this isomorphism, $\chi_{2n} \to \chi^n$, where $\chi^n = \chi \vee \ldots \vee \chi$, the cup product n times.

For all $n \geqslant 0$, we have a G-module isomorphism

$$\mathfrak{f}\mathfrak{r}^n/\mathfrak{r}^{n+1} \xrightarrow{\sim} \mathfrak{g} \otimes \underbrace{R/R' \otimes \ldots \otimes R/R'}_{n}\ ,$$

where G acts diagonally on the right (proof!). Hence we obtain an appropriate isomorphism of cohomology in dimension 2n+1 and under this, $\chi_{2n+1} \to \chi_1 \chi^n = \chi_1 \vee \chi^n$.

<u>THEOREM 1. cd G $\leqslant$ 2n-1 if, and only if, $\chi^n = 0$ $(n \geqslant 1)$;</u>
<u>cd G $\leqslant$ 2n if, and only if, $\chi_1 \chi^n = 0$ $(n \geqslant 0)$.</u>

Thus G has finite cohomological dimension if, and only if, χ is nilpotent under the cup product multiplication. (Compare this with the Corollary on p.418 of Serre [23].)

8.3 Some groups of cohomological dimension 2.

Groups of cohomological dimension 2 are by no means rare. Indeed, the difficulty lies in distinguishing any common group-theoretic features among these groups.

Perhaps the simplest example is the additive group of rationals:

<u>PROPOSITION 4. If G is torsion free abelian of rank 1 and not finitely generated (i.e., not cyclic) then</u>

<u>(i) cd G = 2</u>

<u>and (ii) $H_2(G,A) = 0$ for all A.</u>

<u>Proof</u>. Since G is countable, let $g_0, g_1, \ldots$ be a set of generators of G. If $\langle h_i \rangle = \langle g_0, \ldots, g_i \rangle$, then $h_0, h_1, \ldots$ also generate G and $\langle h_0 \rangle \leqslant \langle h_1 \rangle \leqslant \ldots$, so that

$$h_{i+1}^{n_{i+1}} = h_i$$

for all $i \geqslant 0$ and some n_i.

Choose F free on $x_0, x_1, \ldots$ and let π: $F \to G$ be the homomorphism extending $x_i \mapsto h_i$. If $R = \operatorname{Ker}\pi$, R is the normal closure of elements $y_1, y_2, \ldots$, where

$$y_i = x_{i-1}^{-1} x_i^{n_i} \quad , \text{ all } i \geqslant 1.$$

We assert

$$0 \to \mathfrak{r}/\mathfrak{f}\mathfrak{r} \to \mathfrak{f}/\mathfrak{f}\mathfrak{r} \to \mathbb{Z}G \to \mathbb{Z} \to 0 \qquad (1)$$

is a free resolution. This will follow if we prove $\mathfrak{r}/\mathfrak{f}\mathfrak{r}$ is free on all $(1-y_i) + \mathfrak{f}\mathfrak{r}$. These elements do generate $\mathfrak{r}/\mathfrak{f}\mathfrak{r}$ as G-module: cf. Remark (7), p.37. Suppose

$$\Sigma(1-y_i)\alpha_i \in \mathfrak{f}\mathfrak{r}$$

with not all α_i in $\mathfrak{f}\mathfrak{r}$. Now

$$1-y_i \equiv -(1-x_{i-1}) + (1-x_i^{n_i}) \qquad (\text{mod } \mathfrak{f}\mathfrak{r})$$

$$\equiv -(1-x_{i-1}) + (1-x_i)\xi_i \qquad (\text{mod } \mathfrak{f}\mathfrak{r}), \qquad (2)$$

where $\xi_i = 1 + x_i + \ldots + x_i^{n_i-1}$. So

$$-\sum_i(1-x_{i-1})\alpha_i + \sum_i(1-x_i)\xi_i\alpha_i \in \mathfrak{f}\mathfrak{r}.$$

Let k be the smallest suffix for which α_i is <u>not</u> in $\mathfrak{f}\mathfrak{r}$. Since

$\mathfrak{f}/\mathfrak{f}\mathfrak{r}$ is free on the $(1-x_i) + \mathfrak{f}\mathfrak{r}$, the coefficient of $(1-x_{k-1}) + \mathfrak{f}\mathfrak{r}$ must be in $\mathfrak{f}\mathfrak{r}$, i.e., $\alpha_k \in \mathfrak{f}\mathfrak{r}$ - a contradiction.

Thus cd $G \leqslant 2$. If we had cd $G = 1$ and $1 \to B \to A \to G \to 1$ is exact with A free abelian, then the sequence splits and so G is free abelian. Hence G is cyclic. Thus cd $G = 2$.

We have now proved (i). To see (ii), we take any left module A and perform $\underset{ZG}{\otimes} A$ on (1). We assert

$$\mathfrak{r}/\mathfrak{f}\mathfrak{r} \underset{ZG}{\otimes} A \to \mathfrak{f}/\mathfrak{f}\mathfrak{r} \underset{ZG}{\otimes} A$$

is injective. This will prove (ii). Now

$$\sum_{i\geqslant 1} ((1-y_i) + \mathfrak{f}\mathfrak{r}) \otimes a_i \to 0$$

implies (by (2) above)

$$-\sum_i ((1-x_{i-1}) + \mathfrak{f}\mathfrak{r}) \otimes a_i + \sum_i ((1-x_i) + \mathfrak{f}\mathfrak{r}) \otimes \xi_i^{\pi} a_i = 0. \tag{3}$$

Since $\{(1-x_i) + \mathfrak{f}\mathfrak{r}\}$ freely G-generates $\mathfrak{f}/\mathfrak{f}\mathfrak{r}$, (3) implies $\xi_i^{\pi} a_i = a_{i+1}$, all $i \geqslant 1$ and $0 = a_1$. Hence $a_i = 0$ all i, as required.

<u>Remark</u>: Compare part (ii) with the exercise at the end of §3.6 (p.47).

The two most important classes of groups of cohomological dimension $\leqslant 2$ are (i) torsion free groups with one defining relation and (ii) knot groups. We shall discuss the first class in more detail in the next section. Concerning the second class we limit ourselves here to a brief description. (Cf. [7] for a nice introduction to knot theory.)

A subset K of $\mathbb{R}^3$ is called a <u>knot</u> if it is the image of a homeomorphism of the unit circle. Then $\mathbb{R}^3 - K$ is pathwise connected and the fundamental group of $\mathbb{R}^3 - K$ is called <u>the group of K</u> and written $\pi(K)$.

Knots K_1, K_2 are called <u>equivalent</u> if there exists a homeomorphism $f\colon \mathbb{R}^3 \to \mathbb{R}^3$ such that $K_1 f = K_2$. A knot is <u>polygonal</u> if K is the union of a finite number of closed straight line segments. If K is equivalent to a polygonal knot, K is called <u>tame</u> (otherwise <u>wild</u>).

The tame knot K is <u>trivial</u> if it is equivalent to the circle in $\mathbb{R}^3$. Then its group is infinite cyclic.

The following hard topological theorem was proved by Papakyriakopoulos in 1957 [21]:

<u>PAPAKYRIAKOPOULOS' THEOREM. If K is a tame knot, then $\operatorname{cd} \pi(K) \leqslant 2$. Moreover, K is trivial if, and only if, $\pi(K)$ is infinite cyclic</u>.

We shall see below that certain constructions preserve or bound the cohomological dimension: notably free products (§8.6) and countable direct limits (§8.5). Whenever this happens, we can produce new groups of dimension $\leqslant 2$. There are examples of this in §8.5.

8.4 One relator groups.

THEOREM 2. (Lyndon [13].) Let $1 \to R \to F \xrightarrow{\pi} G \to 1$ be a free presentation, where R is the normal closure of the single element w^h $(h \geqslant 1)$ and w is not itself a proper power of any element in F. If $C = \langle w^\pi \rangle$, then

$$H^q(G,A) \cong H^q(C,A) \text{ and } H_q(G,A) \cong H_q(C,A)$$

for all G-modules A and all $q \geqslant 3$.

COROLLARY 1. If $h = 1$ (i.e., $C = 1$), then $\operatorname{cd} G \leqslant 2$.

COROLLARY 2. If G has one defining relation, then G is torsion-free if, and only if, the relation is not a proper power. (Use Corollary 1 and Proposition 2, (ii), p.121.)

A non-homological proof of Corollary 2 is given by Karrass, Magnus and Solitar in [11]. Their methods yield more: if G is not torsion-free, then every element of finite order lies in some conjugate of C. (Cf. also [15], pp. 266, 269.)

The hard and purely group theoretic part of Theorem 2 is

LYNDON'S IDENTITY THEOREM. ([13], or better [14].) In the notation of Theorem 2, there is a G-module isomorphism:

$$R/R' \cong \mathbb{Z}(G/C),$$

where G/C is the (right) coset space and $\mathbb{Z}(G/C)$ is the free *abelian group on G/C made into a G-module via the permutational representation of G on G/C.*

Proof of Theorem 2. Let $E = \langle w \rangle$ and T be a right transversal of ER in F. If M is G-free on the single generator m, let

$$\mu: \ M \longrightarrow \mathfrak{f}/\mathfrak{f}\mathfrak{r}$$

be $m \longmapsto (1-w^h) + \mathfrak{f}\mathfrak{r}$. Clearly $M\mu = \mathfrak{r}/\mathfrak{f}\mathfrak{r}$. Suppose $m(\Sigma \alpha_i t_i)^\pi$ is in Ker μ (where $\alpha_i \in \mathbb{Z}E$, $t_i \in T$). Then

$$\Sigma\, \alpha_i (1-w^h) t_i \in \mathfrak{f}\mathfrak{r}$$

(using also the commutativity of E). By the Lyndon Identity Theorem, $\mathfrak{r}/\mathfrak{f}\mathfrak{r}$ is $\mathbb{Z}$-free on all $(1-w^h)t + \mathfrak{f}\mathfrak{r}$ and so $\alpha_i \in \mathfrak{f}$, all i. Hence $\alpha_i \in \mathfrak{f} \cap \mathbb{Z}E$ = augmentation ideal of E. Thus Ker $\mu = m(1-w^\pi)\mathbb{Z}G$. Note that if $h = 1$, then μ is injective.

We now have a G-free resolution of $\mathbb{Z}$:

$$\ldots \xrightarrow{\sigma} M \xrightarrow{\tau} M \xrightarrow{\sigma} M \xrightarrow{\mu} \mathfrak{f}/\mathfrak{f}\mathfrak{r} \longrightarrow \mathbb{Z}G \longrightarrow \mathbb{Z} \longrightarrow 0, \qquad (1)$$

where σ: $m \longmapsto m(1-w^\pi)$ and τ: $m \longmapsto m(1 + w^\pi + \ldots (w^\pi)^{h-1})$.

From dimension 2 and up this resolution is simply the blown up version of the usual resolution for the cyclic group C (i.e., $* \otimes_{\mathbb{Z}C} \mathbb{Z}G$) (§3.3, p.39). Moreover, there is a natural isomorphism

$$\mathrm{Hom}_G(* \otimes_{\mathbb{Z}C} \mathbb{Z}G, -) \cong \mathrm{Hom}_C(* , -),$$

where * places a C-module and - places a G-module.

Hence G and C have the same cohomology in all dimensions > 2. For similar reasons they have the same homology.

Remark. If $\mathfrak{a}$ is the right ideal in $\mathbb{Z}F$ on $1-w^h$, then $M \cong \mathfrak{a}/\mathfrak{a}\mathfrak{r}$. Hence (1) extends the sequence (*) of p.37.

Exercises.

1. If H is a subgroup of G and (P_i) is a projective resolution of $\mathbb{Z}$ over H, then $(P_i \otimes_H G)$ is a projective resolution of $\mathbb{Z} \otimes_H G$ over G. (We write $M \otimes_H G$ for $M \otimes_{\mathbb{Z}H} \mathbb{Z}G$.)

2. Let H be a subgroup of G and (P_i), (X_i) projective resolutions of $\mathbb{Z}$ over H, G, respectively. Assume

$$\mathrm{Ker}(P_k \to P_{k-1}) = K, \quad \mathrm{Im}(X_k \to X_{k-1}) = Y$$

are such that $\mathrm{Coker}\,(K \otimes_H G \to P_k \otimes_H G) \cong Y$. Prove that $H^q(G,A) \cong H^q(H,A)$ for all G-modules A and all $q > k$.

3. (i) Prove that a free abelian group of rank 2 has cd = 2.

(ii) If A, B are free abelian of rank 2, then A*B has cd 2 (cf. §8.6, below). Prove however that A*B is not a one relator group (i.e., has no presentation with one relation). (Hint: Use the Gruško-Neumann theorem [12], p.59, and the Freiheitssatz, [15], p.252.)

8.5 Direct limits.

PROPOSITION 5. (Berstein [4].) If $G = \varinjlim G_i$, where I is countable, then $\mathrm{cd}\, G \leqslant 1 + \sup\{\mathrm{cd}\, G_i;\ i \in I\}$.

Some preliminary remarks first. Let $(A_i; \alpha_{ij})_I$ be a direct system of modules over the direct system of groups $(G_i; \gamma_{ij})_I$: this means that each A_i is a G_i-module and for $i \leqslant j$, a in A_i, g in G_i,

$$(ag)\alpha_{ij} = (a\alpha_{ij})(g\gamma_{ij}).$$

Then $A = \varinjlim A_i$ is, in a natural way, a module over $G = \varinjlim G_i$. (For details about these and other relevant matters see, e.g., [5], chapter 2, §6, nos. 6,7.)

Let $B_i = A_i \underset{\mathbb{Z}G_i}{\otimes} \mathbb{Z}G$ and $\beta_{ij}\colon B_i \to B_j$ be $a \underset{G_i}{\otimes} x \to a\alpha_{ij} \underset{G_j}{\otimes} x$. It follows that $(B_i; \beta_{ij})_I$ is a direct system of G-modules.

LEMMA 2. $\varinjlim B_i \cong A$.

Proof. Let $\alpha_i : A_i \to A$. Then the mappings $B_i \to A$ given by $a \otimes x \mapsto (a\alpha_i)x$ obviously yield a module epimorphism $\beta\colon \varinjlim B_i \to A$.

Suppose $b\beta = 0$ and let b be the image of $y \in B_i$. Now y has the form

$$y = \sum_{r=1}^{s} y_r \otimes t_r ,$$

with $y_1,\dots,y_s$ in A_i and $t_1,\dots,t_s$ in a right transversal of the image of G_i in G. Choose $j \geqslant i$ so that $t_1,\dots,t_s$ belong to the image of G_j in G and let $t'_r \to t_r$ under $G_j \to G$. Then

$$y\beta_{ij} = a \otimes 1, \text{ where } a = \Sigma\,(y_r\alpha_{ij})t'_r \in A_j .$$

But $(B_j \to A) = (B_j \to B \to A)$ and so $a\alpha_j = 0$. Hence $a\alpha_{jk} = 0$ for some $k \geqslant j$ so that $y\beta_{ik} = 0$. Thus $b = 0$ and β is one-one.

<u>Proof of Proposition 5</u>. For each i in I we construct a G_i-free resolution of $\mathbb{Z}$ as follows: Let X_{io} be the free G_i-module on the set $\mathbb{Z}$ (i.e., qua abstract set) and ϵ: $X_{io} \to \mathbb{Z}$ the obvious epimorphism; then X_{i1} is to be the free G_i-module on the set $\operatorname{Ker} \epsilon$ and d_{i1}: $X_{i1} \to X_{io}$ the homomorphism $X_{i1} \to \operatorname{Ker} \epsilon \hookrightarrow X_{io}$; and so on.

We now have a direct system of resolutions:

$$\begin{array}{ccccc} \cdots\to X_{in} & & \longrightarrow & & X_{i,n-1} \to \cdots \to X_{io} \to \mathbb{Z} \to 0, \\ & \searrow & & \nearrow & \\ & & K_{in} & & \\ & \nearrow & & \searrow & \\ 0 & & & & 0 \end{array}$$

where $K_{in} = \operatorname{Ker} d_{i,n-1}$. Taking the direct limit preserves exactness and moreover, $X_n = \varinjlim_i X_{in}$ is G-free on $K_n = \varinjlim_i K_{in}$.

So we obtain a G-free resolution of $\mathbb{Z}$:

$$\begin{array}{ccccc} \cdots\to X_n & & \longrightarrow & & X_{n-1} \to \cdots \to X_0 \to \mathbb{Z} \to 0. \\ & \searrow & & \nearrow & \\ & & K_n & & \\ & \nearrow & & \searrow & \\ 0 & & & & 0 \end{array}$$

Suppose $\sup\{\operatorname{cd} G_i\} = k$. By Proposition 1, (i)⇒(iii), K_{ik} is G_i-projective, for each i. So all we need prove is the existence of a short exact sequence

$$0 \to Q \to P \to K_k \to 0$$

with P, Q both G-projective.

By Lemma 2, $K_k \cong \varinjlim L_i$, where $L_i = K_{ik} \otimes_{\mathbb{Z}G_i} \mathbb{Z}G$ and of course L_i is G-projective.

We now (for the first time) use the countability of I. This enables us to pick a cofinal sequence $s_1 \leqslant s_2 \leqslant \ldots$ in I. Let $M_i = L_{s_i}$ and $\lambda_i : M_i \to M_{i+1}$. Then $\varinjlim L_i \cong \varinjlim M_i$ and

$$\varinjlim M_i = \coprod M_i/M,$$

where $M = \langle x - x\lambda_i; \ \ x \in M_i, \ i \geqslant 1\rangle$. The direct sum $\coprod M_i$ is G-projective and

$$0 \to \coprod M_i \xrightarrow{\lambda} \coprod M_i \xrightarrow{\pi} \varinjlim M_i \to 0$$

is exact, where π is the natural projection and

$$\lambda : \ x \mapsto x - x\lambda_i \quad \text{for } x \in M_i.$$

<u>Remark</u>. Barbara Osofsky [20] has recently used Berstein's result to obtain the following generalisation: <u>If I in Proposition 5 has cardinal $\aleph_n$, then $\operatorname{cd} G \leqslant 1 + n + m$, where $m = \sup\{\operatorname{cd} G_i; \ i \in I\}$.</u>

COROLLARY 1. *Let $1 = H_0 \leqslant H_1 \leqslant \ldots$ be a series of normal subgroups of G so that $\operatorname{cd} G/H_i \leqslant k$ for all $i \geqslant 0$. Then $\operatorname{cd} G/\bigcup H_i \leqslant 1+k$.*

COROLLARY 2. *If G is countable and locally of $\operatorname{cd} \leqslant k$, then $\operatorname{cd} G \leqslant 1+k$.*

Corollary 2 gives another immediate proof that torsion-free abelian groups of rank 1 have $\operatorname{cd} \leqslant 2$ (Proposition 4). It also implies that *every countable locally free group has $\operatorname{cd} \leqslant 2$.* Osofsky's generalisation of Proposition 5 raises the interesting possibility that there exist locally free groups of cd precisely n, for every positive integer n.

We conclude this section with two instructive examples of countable, locally free groups of cd = 2.

Example 1. Let $F_1, F_2, \ldots$ be a sequence of free groups, each of countably infinite rank. Choose some fixed prime p and any isomorphism

$$\alpha_n : A_n \xrightarrow{\sim} (A_{n+1})^p,$$

where $A_n = F_n/F_n'$. Let $\varphi_n : F_n \to F_{n+1}$ be any homomorphism so that

$$\begin{array}{ccc} F_n & \xrightarrow{\varphi_n} & F_{n+1} \\ \downarrow & & \downarrow \\ A_n & \xrightarrow{\alpha_n} & A_{n+1} \end{array}$$

commutes. Form $G = \varinjlim (F_n;\varphi_n)$, the direct limit with respect to the homomorphisms φ_n. By Corollary 2, cd $G \leqslant 2$ and we assert cd $G \nless 2$. This follows from two facts: (i) $G/G' \cong \varinjlim F_n/F_n'$ and the right hand side is a non-trivial p-divisible group (so that $G^pG' = G$); and (ii) the following result:

LEMMA 3. If $H^2(G,A) = 0$ for all trivial G-modules A, then G/G' is a (possibly trivial) free abelian group.

Proof. The universal coefficient theorem (p. 49) implies $H_1(G,\mathbb{Z})$ is $\mathbb{Z}$-free, i.e., G/G' is free.

Alternatively, if $1 \to R \to F \to G \to 1$ is exact with F free and $\tilde{R} = R/[R,F]$, $\tilde{F} = F/[R,F]$, then $1 \to \tilde{R} \to \tilde{F} \to G \to 1$ splits, i.e., $\tilde{F} \cong G[\tilde{R}$. Hence $\tilde{F}_2 \cong G_2$ and therefore $F/F_2 \cong G/G_2 \times \tilde{R}$. Since F/F_2 is $\mathbb{Z}$-free, so is G/G_2.

Example 2. Let F_n be as in example 1 and choose any isomorphism ψ_n: $F_n \rightrightarrows (F_{n+1})'$. Then $G = \varinjlim (F_n;\ \psi_n)$ is now a perfect group and we still have cd $G = 2$. This last is a consequence of the Stallings-Swan result that cd $G = 1$ implies G is free. However we may avoid using this difficult theorem here by the following argument, due also to Swan.

PROPOSITION 6. If G is locally free and cd $G \leqslant 1$, then G is not perfect.

Proof. We know $\mathfrak{g}$ is projective. Let $\{x_i;\ i \in I\}$ generate G and let M be G-free on $\{e_i;\ i \in I\}$. Map π: $M \to \mathfrak{g}$ by $e_i \mapsto 1-x_i$. So there exists $\tau: \mathfrak{g} \to M$ with $\tau\pi = 1_{\mathfrak{g}}$. Let

$$(1-x_j)\tau = \sum_i e_i a_{ij}, \quad a_{ij} \in \mathbb{Z}G \tag{1}$$

and apply $\pi\tau$:

$$\sum_i e_i a_{ij} = \sum_{k,i} e_k a_{ki} a_{ij},$$

and therefore

$$a_{kj} = \sum_i a_{ki} a_{ij} . \tag{2}$$

Assume G is perfect. Then $\mathfrak{g} = \mathfrak{g}^2$ and so, since $M = \mathfrak{g}\tau \oplus N$, for some N, $M\mathfrak{g} = \mathfrak{g}\tau \oplus N$, whence

$$\sum e_i a_{ij} = (1-x_j)\tau \in M\mathfrak{g} .$$

Since $M/M\mathfrak{g}$ is $\mathbb{Z}$-free on the $e_i + M\mathfrak{g}$, therefore

$$\text{each } a_{ij} \in \mathfrak{g}. \tag{3}$$

As $\tau \neq 0$, there exists j such that $(1-x_j)\tau \neq 0$: say $j = 1$. Then not all a_{k1} are zero; but only a finite number in (1) are $\neq 0$: say $a_{k1} \neq 0$ for $k = 1,\dots,n$. In (2),

$$a_{k1} = \sum_i a_{ki} a_{i1}$$

shows (since $a_{k1} = 0$ if $k \neq 1,\dots,n$) that we can restrict to $i = 1,\dots,n$. Let the group elements involved in a_{ki}, $k,i = 1,\dots,n$ generate the finitely generated subgroup H of G. Let $\mathfrak{a}$ be the <u>left</u> ideal generated by a_{k1}, $k = 1,\dots n$. Now for all $1 \leqslant i, j \leqslant n$,

$$a_{ij} \in \mathfrak{g} \cap \mathbb{Z}H = \mathfrak{h}$$

by (3); and $a_{kl} \in \mathfrak{f}\mathfrak{a}$ by (2). Hence $\mathfrak{a} \leqslant \mathfrak{f}\mathfrak{a}$: i.e., $\mathfrak{a} = \mathfrak{f}\mathfrak{a} = \mathfrak{f}^n\mathfrak{a}$, all n, and so $\mathfrak{a} \leqslant \mathfrak{f}^n$, all n. But $\mathfrak{a} \neq 0$ as not all a_{kl} were zero. Consequently $\bigcap \mathfrak{f}^n \neq 0$. But augmentation ideals of free groups are residually nilpotent (p.56). This is the required contradiction.

8.6 Free products.

(Notation: If $H \to G$, then $M \underset{H}{\otimes} G$ shall mean $M \underset{\mathbb{Z}H}{\otimes} \mathbb{Z}G$.)

Recall that the coproduct in the category of groups is $*$, the free product; and that the coproduct in the category of modules over a ring is $\oplus$, the direct sum. We usually denote coproducts by the neutral symbol $\coprod$.

(Cf. [17] for the definitions of the various categorical notions used in this section: products, pull-back, equalizers; and their co-mates.)

THEOREM 3. If $G = \underset{i \in I}{*} G_i$, then for all $q \geqslant 2$ and all G-modules A,

$$H^q(G,A) \cong \prod_i H^q(G_i,A).$$

COROLLARY. $\operatorname{cd} G = \sup\{\operatorname{cd} G_i \; ; \; i \in I\}$.

Theorem 3 is false for $q = 1$. For example, let F be free on x_1, x_2 and A be an infinite cyclic group on which x_1, x_2 act non-trivially: $ax_i = -a$. Then $H^1(\langle x_i\rangle, A)$ is cyclic of order 2 but $H^1(F,A) \cong \mathbb{Z} \oplus (\mathbb{Z}/2\mathbb{Z})$.

For the proof of Theorem 3 we need a simple general observation (Lemma 4) and a fact about the augmentation ideal of a free product (Proposition 7).

LEMMA 4. If $\dots \to P_{i2} \to P_{i1} \to \mathfrak{g}_i \to 0$ is a G_i-projective resolution of $\mathfrak{g}_i$ and we write $\bar{M} = M \otimes_{G_i} G$ for any G-module M, then

$$\dots \to \coprod_i \bar{P}_{i2} \to \coprod_i \bar{P}_{i1} \to \coprod_i \bar{\mathfrak{g}}_i \to 0$$

is a G-projective resolution of $\coprod \bar{\mathfrak{g}}_i$.

Proof = exercise.

LEMMA 5. Res : $\mathrm{Der}(G,A) \stackrel{\sim}{\to} \prod_i \mathrm{Der}(G_i,A)$.

Proof. Clearly the restriction Res does give an additive homomorphism and this is injective.

Suppose (δ_i) is a family in the right hand side. Each δ_i gives a homomorphism $d_i : G_i \to G[A$, viz., $gd_i = (g, g\delta_i)$ (cf. p.32). Hence, by the definition of free products, there exists a homomorphism $d: G \to G[A$ extending the d_i. Then d gives the derivation $G \to A$ whose restriction is (δ_i).

<u>PROPOSITION 7. If $G = \underset{i\in I}{*} G_i$, then</u>

$$\underline{\mathfrak{g} \cong \coprod_I \mathfrak{g}_i \underset{G_i}{\otimes} G .}$$

<u>Proof</u>. Use Lemma 5 and the definition of direct sum.

<u>Proof of Theorem 3</u>. By Lemma 4 and Proposition 7, we have a G-projective resolution $(\coprod_I F_{in})_{n\geq 1}$ of $\mathfrak{g}$. Hence

$$\begin{array}{ccccccc} \cdots \to \coprod_I F_{i1} & & \longrightarrow & & \mathbb{Z}G \to \mathbb{Z} \to 0 \\ & \searrow & & \nearrow & \\ & & \mathfrak{g} & & \\ & \nearrow & & \searrow & \\ 0 & & & & 0 \end{array}$$

is a G-projective resolution of $\mathbb{Z}$. The result now follows since

$$\mathrm{Hom}_G(\coprod_I F_{in}, M) \cong \prod_i \mathrm{Hom}_{G_i}(P_{in}, M)$$

is a natural isomorphism.

We may express Proposition 7 in a prettier way. Let Mod be the category of "all modules over all groups": the objects are pairs (G,M), where G is a group and M is a G-module; and a morphism: $(G,M) \to (H,N)$ is a pair (φ,μ), where $\varphi: G \to H$ is a group homomorphism and $\mu: M \to N$ is a G-module homomorphism (i.e., $(mg)\mu = (m\mu)(g\varphi)$). If (G_i, M_i) is a family in Mod, and we let $G = * G_i$, then

$$\coprod (G_i, M_i) = (G, \coprod M_i \underset{G_i}{\otimes} G).$$

If Gps is the category of all groups and Aug: Gps $\longrightarrow$ Mod is the functor $G \longrightarrow (G, \mathfrak{g})$, then Proposition 7 asserts that <u>Aug preserves coproducts</u>.

<u>Exercises</u>.

1. What do products look like in Mod ? Does Aug preserve products?

2. Prove that Aug is the left adjoint of $(G,M) \longrightarrow G[M$. (This implies that Aug preserves all colimits.)

Proposition 7 and Theorem 3 may be generalised. Let γ_i: $H \longrightarrow G_i$ be a family of group homomorphisms and E the corresponding push-out (or fibre-coproduct, or cocartesian). In the category of groups this is the free product with amalgamation.

Note that if $G = * G_i$, then $G \longrightarrow E$ is the coequalizer of $(\gamma_i' : H \longrightarrow G)$, where γ_i' is $H \longrightarrow G_i \hookrightarrow G$. Hence we may prove exactly as in Lemma 5 that

$$\mathrm{Der}(E,A) \xrightarrow{\mathrm{Res}} \prod \mathrm{Der}(G_i,A) \underset{\vdots}{\overset{\delta_i}{\rightrightarrows}} \mathrm{Der}\,(H,A)$$

is exact, where $\delta_i = \mathrm{Der}(\gamma_i,A)$: i.e., Res is the equalizer of the family ($\mathrm{Der}(\gamma_i,A)$). From this we deduce immediately

PROPOSITION 8. *The augmentation ideal $\mathfrak{e}$ of E is the push-out in Mod_E of the family $(\gamma_i \otimes 1 : \mathfrak{h} \otimes_H E \longrightarrow \mathfrak{g}_i \otimes_{G_i} E)$.*

<u>Exercise</u>. Let D be a diagram in Mod and D' the corresponding

diagram in Gps (i.e., (G_i, M_i) contributes G_i). Put $G = \operatorname{colim} D'$ and let D'' be the diagram in Mod_G determined by D : i.e., (G_i, M_i) contributes $M_i \underset{G_i}{\otimes} G$. Put $M = \operatorname{colim} D''$. Then $(G,M) = \operatorname{colim} D$.
(In view of exercise 2 above, this fact contains Proposition 8.)

We may transform Proposition 8 into a statement concerning an exact sequence. Take $k \in I$ and for each $i \neq k$, let M_i be an isomorphic copy of $\mathcal{J} \underset{H}{\otimes} E$. Then define

$$\mu_i\colon \; M_i \to \coprod \mathcal{J}_i \underset{G_i}{\otimes} E \text{ to be } a \mapsto a(\overline{\gamma}_i - \overline{\gamma}_k)$$

($a \in \mathcal{J} \underset{H}{\otimes} E$, $\overline{\gamma}_i = \gamma_i \otimes 1$).

We now have the exact sequence of E-modules

$$\coprod_{i \neq k} M_i \xrightarrow{\mu} \coprod \mathcal{J}_i \underset{G_i}{\otimes} E \to \mathfrak{U} \to 0. \qquad (*)$$

Moreover, if each γ_i is injective, so is μ.

Assume now that each γ_i is injective. Choose E-projective resolutions (X_n) of $\coprod M_i$ and (Y_n) of $\mathfrak{U}$. Then by (*), $(Z_n) = (X_n \oplus Y_n)$ gives an E-projective resolution of $\coprod \mathcal{J}_i \underset{G_i}{\otimes} E$ and we have the exact sequence of E-complexes:

$$0 \to (X_n) \to (Z_n) \to (Y_n) \to 0.$$

(cf. Cartan-Eilenberg, p.80.) For any E-module A,

$$0 \to \mathrm{Hom}_E(Y_n, A) \to \mathrm{Hom}_E(Z_n, A) \to \mathrm{Hom}_E(X_n, A) \to 0$$

is also exact (as X_n is a direct summand of Z_n). The corresponding homology sequence (cf. pp. 19,20) now yields (cf. the proof of Theorem 3)

<u>THEOREM 4. (Barr-Beck [2], p.310) If E is the push-out of the monomorphic family ($\gamma_i : H \to G_i$; $i \in I$), then the following sequence is exact:</u>

$$0 \to \mathrm{Der}(E,A) \to \prod_{i\in I} \mathrm{Der}(G_i,A) \to \mathrm{Der}(H,A)^{I'} \to H^2(E,A) \to \prod_{i\in I} H^2(G_i,A) \to H^2(H,A)^{I'} \to H^3(E,A) \to \ldots$$

(Here M^J is the product of J copies of M and I' is the set I with one element removed.)

<u>COROLLARY. Let $k = \sup\{\mathrm{cd}\, G_i \; ; \; i\in I\}$.</u>

<u>(i) If cd H < k, then cd E = k.</u>

<u>(ii) If cd H = k, then cd E equals k or 1+k.</u>

<u>Remarks.</u>

(1) The condition that each γ_i is injective is necessary. For otherwise the corollary would apply to all push-outs and in particular to all coequalizers. But this is obviously false.

(2) The maps $H^q(H,A)^{I'} \to H^{q+1}(E,A)$ are not zero in general: if they were, the corollary would read $\mathrm{cd}E = \sup\{\mathrm{cd}\, G_i\}$.

Counterexample: Let H, G_1, G_2 be free cyclic on x, y_1, y_2, respectively and γ_i: $x \mapsto y_i^2$. Then $A = \langle y_1^2, y_1y_2\rangle$ is abelian (as y_1^2 is central in E) and is not cyclic. So cd A = 2.

(3) The above counterexample also shows that cd E = 1+k can occur in case (ii) of the corollary. So can cd E = k: Let H be free on x, G_1 free on a,b and G_2 free on c and suppose $x\gamma_1 = a$, $x\gamma_2 = c$. Then E is free on b,c.

Exercises. (1-4 need an elementary knowledge of the Ext functor.)

1. Let M be a G-module and $0 \to M \to Q_0 \to Q_1 \to \cdots$ an injective resolution of M. Prove that the homology of the complex $(\mathrm{Der}(G,Q_i))$ is $(\mathrm{Der}(G,M), H^2(G,M), H^3(G,M), \ldots)$.

2. If $H \leqslant G$ and M is an injective G-module, prove M is H-injective.

3. Let E be the push-out of the monomorphic pair: $H \to A$, $H \to B$. If Q is E-injective, prove that

$$0 \to \mathrm{Der}(E,Q) \xrightarrow{\mathrm{Res}} \mathrm{Der}(A,Q) \oplus \mathrm{Der}(B,Q) \xrightarrow{\rho} \mathrm{Der}(H,Q) \to 0$$

is exact, where ρ: $(\delta', \delta'') \mapsto$ restriction to H of $\delta' - \delta''$.

4. Deduce Theorem 4 (for the case $|I| = 2$) from exercises 1 and 3.

5. If $E = \langle x_1,\ldots,x_m, a_1, b_1,\ldots,a_n, b_n \mid x_i^{s_i} = 1$ $(i=1,\ldots,m)$, $x_1\cdot\ldots\cdot x_m \Pi[a_i,b_i] = 1\rangle$, prove that, for all $q \geqslant 3$ and all E-modules A,

$$H^q(E,A) \cong H^q(\langle x_1\rangle, A)\oplus\ldots\oplus H^q(\langle x_m\rangle, A).$$

(E is a <u>Fuchsian group</u> with compact orbit space and genus n.)

<u>Problem</u>. Find the class of all diagrams D in groups so that cd colim D is finite whenever sup{cd D_i; all vertices i} is finite.

8.7 Extensions.

<u>PROPOSITION 9. Suppose $S \lhd G$ and cd S = s and cd G/S = q.</u> Then
<u>(i) cd $G \leq s+q$ and</u>
<u>(ii) $H^{s+q}(G,A) \cong H^q(\, G/S,\, H^s(S,A)\,)$ for every G-module A.</u>

We shall omit the proof. It involves a consideration of the spectral sequence associated with a group extension. The result is frequently called the <u>"maximum principle for spectral sequences"</u>.

In order to use Proposition 9, however, we must explain how the abelian groups $H^p(S,A)$ are given the structure of G/S-modules.

Let S be a (not necessarily normal) subgroup of G, g an element in G and $S^g = g^{-1}Sg$. Let ρ be the usual restriction $Mod_G \to Mod_S$ (cf. p.88, bottom) and ρ^g the corresponding restriction $Mod_G \to Mod_{S^g}$.

Now F: $A \to (A\rho)^S$ is a functor from Mod_G to abelian groups and F is extended by the minimal cohomological functor $H^*(S,-\rho)$. (For the proof of minimality see p.90: the argument there had nothing to do with H being of finite index in G.)

Similarly F^g: $A \to (A\rho^g)^{S^g}$ is extended by the minimal cohomological functor $H^*(S^g, -\rho^g)$.

Let $\tilde{g}_A$: $(A\rho)^S \to (A\rho^g)^{S^g}$ be $a \mapsto ag$. Clearly $\tilde{g}$ is a natural isomorphism: $F \to F^g$. By Theorem 1, Chapter 6 (p.86), there exists one and only one natural homomorphism

$$H^*(S, -\rho) \to H^*(S^g, -\rho^g)$$

extending $\tilde{g}$. Write this also as $\tilde{g}$. This is called <u>conjugation</u> by g.

In particular, if $S \lhd G$, $\tilde{g}$ is a natural automorphism of $H^*(S, \ )$ (we now omit ρ as being understood). Clearly $\tilde{g}_1\tilde{g}_2 = \widetilde{g_1 g_2}$ and hence, for any A, $g \mapsto \tilde{g}^p_A$ gives a representation of G on $H^p(S,A)$. If $g \in S$, $\tilde{g}^0_A$ is the identity and hence (by uniqueness), $\tilde{g}^p_A$ is the identity for all $p > 0$. We therefore really have a <u>representation of G/S on $H^p(S,A)$</u>. We say that $H^p(S,A)$ is a G/S-module <u>by conjugation</u>. It is now meaningful to construct

$$H^r(G/S, H^p(S,A))$$

for all $p,r \geq 0$.

If we wish to prove that for a particular pair S,G the inequality (i) of Proposition 9 is really an equality, we must find a G-module A so that the right hand side of the isomorphism (ii) is non-zero. This is usually extremely difficult. (Solutions in special cases will occur in the next section.)

Of course, the inequality (i) can quite easily be strict. It can even be strict if G is a product. For example, let $G = Q \times Q$, where Q is the additive group of rationals. Then cd Q = 2 (Proposition 4, §8.3) and cd G = 3 (cf. §8.8).

Note that for any groups $G_1,\ldots,G_n$,

$$n \leqslant \text{cd}(G_1\times\ldots\times G_n) \leqslant \text{cd } G_1 + \ldots + \text{cd } G_n. \quad (*)$$

The left hand inequality comes from the fact that $G_1\times\ldots\times G_n$ contains free abelian subgroups of rank n and these have cd = n (cf. §8.8).

The right hand inequality of (*) yields

PROPOSITION 10. *Let D be any finite diagram in the category of groups and I the set of all vertices. Then*

$$\text{cd lim } D \leqslant \sum_{i\in I} \text{cd } D_i .$$

This is immediate from Proposition 2.9 of [17], p.47.

Another type of result concerning extensions is the following very interesting and useful theorem due to Serre [24] (cf.[23] for the profinite analogue).

SERRE'S EXTENSION THEOREM. If G is a torsion-free group containing a subgroup H of finite index, then cdH = cdG.

A complete proof is given at the end of Swan's paper [32]. Note that Serre's theorem together with the Stallings-Swan theorem shows that if G is torsion-free and contains a free subgroup of finite index then G itself is free.

An interesting special case of Serre's theorem (which was known before from topological considerations) arises if G is torsion-free, finitely generated and has an abelian subgroup A of finite index. In this case, cdG = rank A. Such groups are called Bieberbach groups or crystallographic groups: cf.[34], Chapter 3. Example: If F is a finitely generated free group and R is a normal subgroup of finite index, F/[R,R] is a Bieberbach group.

Exercise. If G is torsion-free and has a locally free normal subgroup H with G/H locally finite, then G is locally free.

8.8 Nilpotent groups.

We shall prove a theorem that describes completely the function cd on the class of all nilpotent groups in terms of a much older, purely group-theoretic function.

Definition. Let G be a polycyclic group (=soluble with maximal condition on subgroups) and suppose $1 = S_0 < S_1 < \ldots < S_k = G$ is a series with cyclic factors S_{i+1}/S_i. The number of infinite factors is an invariant and will be called the <u>Hirsch number</u> of G, h(G). (This was discovered by Hirsch in 1938 [9]; see Scott, p.150 for a modern treatment.)

If G is locally polycyclic, define the Hirsch number h(G) of G to be sup{h(H); all finitely generated subgroups H}.

The Hirsch number is a direct generalisation of the rank of torsion-free abelian groups.

Every finitely generated nilpotent group is polycyclic and therefore every locally nilpotent group has a well-defined Hirsch number.

Recall also that if G is finitely generated nilpotent and torsion-free, then there exists a series from 1 to G with all factors infinite cyclic. (Warning: this is not true for all torsion-free polycyclic groups.)

<u>THEOREM 5. (1) Let G be locally nilpotent. Then cd $G < \infty$ if, and only if, G is torsion-free, nilpotent and has finite Hirsch number</u>.

<u>(2) If G is torsion-free nilpotent and $h(G) < \infty$, then</u>
<u>(i) cd G = h(G) if G is finitely generated;</u>
<u>(ii) cd G = h(G) + 1 if G is not finitely generated.</u>

The special case of Theorem 5 when G is abelian is due to

Varadarajan [33]. Moreover, part (2)(i) for G abelian is essentially the Hilbert syzygy theorem (Cartan-Eilenberg, p.157 or [10], p.45).

We begin with some general facts concerning central extensions. This is a case where the action of G/S on $H^p(S,A)$ can be made sufficiently explicit to cope with the extension problem for Theorem 5.

Consider, for the moment, the general situation of S any subgroup of G and $g \in G$. Let (X_i) be an S-projective resolution of $\mathbb{Z}$ and view this as an acyclic S^g-complex via g^{-1}: $S^g \to S$ (i.e., for u in X, $u.g^{-1}xg = ux$). Let (Y_i) be an S^g-projective resolution of $\mathbb{Z}$ (e.g., (X_i) itself will do). Then there exists a map ξ of S^g-complexes over the identity on $\mathbb{Z}$:

$$\xi:\ Y \to X.$$

(Cf. Cartan-Eilenberg, p.76.) For any $A \in \mathrm{Mod}_G$, let

$$\xi'_A : \mathrm{Hom}_S(X_i,A) \to \mathrm{Hom}_{S^g}(Y_i,A)$$

be $f \mapsto f'$, where $yf' = (y\xi f)g$. Then ξ'_A is a map of complexes and so gives a homomorphism

$$\hat{g}^p_A:\ H^p(S,A) \to H^p(S^g,A).$$

Clearly $\hat{g}^p_A$ is a natural homomorphism for each $p \geqslant 0$. In fact, $\hat{g}$ commutes with connecting homomorphisms (proof!) and so is a natural homomorphism of cohomological functors. Since $\hat{g}^o = \tilde{g}^o$, we conclude $\hat{g} = \tilde{g}$ (Theorem 1, p.86).

As a simple application we have

<u>PROPOSITION 11. If S is central in G and A is a G-module, then g': $a \mapsto ag$ is an S-automorphism of A and $\tilde{g}^p_A = H^p(S, g')$ (i.e., the action of g on $H^p(S,A)$ comes from the coefficient map g').</u>

<u>Proof.</u> We take $Y_i = X_i$ in the above construction and then may set ξ = identity on X (because S is central). Hence $\xi^!_A$ is composition with g'.

<u>COROLLARY. If S is central in G and A is G-trivial, then for all $p \geq 0$, $H^p(S,A)$ is G/S-trivial.</u>

Let F be any functor: $\mathrm{Mod}_G \to \mathfrak{A}$ (= abelian groups) so that for every trivial G-module A we have an isomorphism φ_A: $AF \xrightarrow{\sim} A$ and whenever f: $A \to B$ then $\varphi_A . f = fF.\varphi_B$. We shall then write

$$AF \cong A \text{ for G-trivial } A.$$

This means of course that F is naturally isomorphic (on the trivial G-modules) to the forgetful functor (cf. p.85). We may also say that F on trivial G-modules is <u>representable</u> by Hom($\mathbb{Z}$,).

<u>LEMMA 6. Let C be an infinite cyclic central subgroup of G and consider two conditions on G/C:</u>

<u>(a) cd G/C = q and (b) $H^q(G/C,A) \cong A$ for G/C-trivial A.</u>

<u>Then (a) implies cd G = q+1 and (a) and (b) together imply $H^{q+1}(G,A) \cong A$ for G-trivial A.</u>

<u>Proof</u>. By Proposition 9, cd G $\leq$ q+1 and

$$H^{q+1}(G,A) \cong H^q(G/C,\ H^1(C,A)).$$

Choose $A \in Mod_{G/C}$ so that $H^q(G/C,\ A) \neq 0$. Since A is C-trivial, $H^1(C,A) \cong A$. Both these are G/C-modules and the naturality of the isomorphism together with Proposition 11 imply that the isomorphism is one of G/C-modules. Hence $H^q(G/C,H^1(C,A)) \cong H^q(G/C,A)$ and so cd G = q+1. The second half of the lemma is immediate.

Lemma 6 of course yields part (2)(i) of Theorem 5. It even gives a bit more.

<u>COROLLARY. If G is finitely generated, torsion-free and nilpotent, then</u>

<u>(i) cd G = h(G) = h and</u>

<u>(ii) $H^h(G,A) \cong A$ for G-trivial A.</u>

<u>Exercise.</u> Prove that (ii) fails with G the additive group of rationals and h = 2.

Assume for the moment all of part (2) of Theorem 5 proved. Then the "if" half of part (1) is immediate and the "only if" half is also clear: If G is locally nilpotent and cd G = k, then all finitely generated subgroups have (a) Hirsch number $\leq$ k (by Lemma 6) and hence (b) nilpotence class $\leq$ k. Now (a) implies h(G) $\leq$ k and (b) implies G is nilpotent.

It remains to prove part (2)(ii). Let then G be torsion-free nilpotent, not finitely generated and of Hirsch number h. Then G is countable (induction on the class). By Lemma 6, G is locally of cd $\leq$ h and hence, by Corollary 2 to Proposition 5 (p.135), cd G $\leq$ h+1. So it remains to find a subgroup E of cd exactly h+1.

We proceed by induction on h. When h = 1 the result is true (Proposition 4, p.125). Choose any subgroup S in the centre Z of G so that S has rank 1 and Z/S is torsion-free. Then G/S is also torsion-free and h(G/S) = h-1. If S is finitely generated, G/S is not finitely generated and so cd (G/S) = h, by induction. Hence cd G = h+1 by Lemma 6. If S is not finitely generated, choose E/S to be a finitely generated subgroup of G/S with h(E/S) = h-1. Then cd(E/S) = h-1 and cdE = h+1 by Lemma 7, below.

LEMMA 7. Suppose S is abelian of rank 1, not finitely generated and central in E. If E/S is finitely generated nilpotent and cd E/S = q then cd E = q+2.

Proof. We know cd S = 2 (Proposition 4, p.125) and so cd E $\leq$ q+2.

Also $$H^{q+2}(E,A) \cong H^q(E/S,\ H^2(S,A)).$$

If A is E-trivial, $H^2(S,A)$ is E/S-trivial (Corollary to Proposition 11) and thus the right hand side is $H^2(S,A)$ by the corollary to Lemma 6. By Lemma 3 (§8.5, p. 136), we can find an E-trivial A so that $H^2(S,A) \neq 0$. Then $H^{q+2}(E,A) \neq 0$.

It would be interesting to know how Theorem 5 generalises to soluble groups. Polycyclic groups at least are well-behaved:

<u>LEMMA 8. If G is polycyclic, then cd G < ∞ if, and only if, G is torsion-free and then cd G = h(G).</u>

<u>Proof.</u> Any polycyclic group G has a subgroup H of finite index which is poly-"infinite-cyclic". Now cd H $\leq$ h(H) = h(G) and if G is torsion-free, then Serre's Extension Theorem (p. 148) ensures cd H = cd G. So all will be proved if we can show cd H = h(H).

Let h = h(H) and use an induction on h to prove more: cd H = h(H), and $H^h(H,\mathbb{F}_2) \cong \mathbb{F}_2$. If $K \lhd H$ with H/K infinite cyclic, $H^{h-1}(K,\mathbb{F}_2) \cong \mathbb{F}_2$ as abelian groups, by induction, and so also as H-modules (because there can be no action to worry us!). By Proposition 9,

$$H^h(H,\mathbb{F}_2) \cong H^1(H/K,\mathbb{F}_2) \cong \mathbb{F}_2.$$

One can say something qualitative about soluble groups of finite cohomological dimension. Results of <u>Čarin</u> (On soluble groups of type A_4, Mat.Sb. 52(94)(1960) 895-914; cf. also [16]), together with what we already know, imply the following: <u>The torsion-free soluble group G has finite cohomological dimension if, and only if, there is a nilpotent normal subgroup N of finite Hirsch number and G/N is finitely generated abelian-by-finite.</u>

8.9 Centres.

We have seen how the nilpotent subgroup structure of a group is restricted by the assumption of finite cd. But the centre seems subject to further restrictions. In fact, if G is non-abelian,

$$\mathrm{cd}\ \zeta_1(G) < \mathrm{cd}\ G \qquad (*)$$

in many interesting cases and possibly always. Note that $\mathrm{cd}\ \zeta_1(G)$ can easily equal $\mathrm{cd}\ G - 1$: e.g., $G = A \times F$, where A is free abelian of rank k and F is free of rank 2.

For soluble groups the inequality (*) is easy:

<u>PROPOSITION 12. If $\mathrm{cd}\ G = \mathrm{cd}\ \zeta_1(G)$, then</u>

<u>(i) $G/\zeta_1(G)$ is periodic and $H \cap \zeta_1(G) \neq 1$ for every subgroup $H \neq 1$;</u>

<u>(ii) all soluble subgroups are abelian.</u>

<u>Proof</u>. Write $Z = \zeta_1(G)$ and $\mathrm{cd}\ G = k$.

(i) $A = \langle Z, a\rangle$ is abelian for any a in G and $\mathrm{cd}\ A = k$. Now Z,A are either both finitely generated or both not finitely generated. So they have the same rank (viz. k or k-1) and hence A/Z is finite.

If $H \cap Z = 1$, then H is periodic (because isomorphic to a subgroup of G/Z) and hence H = 1.

(ii) Let S be a soluble subgroup. By (i), SZ/Z is locally finite. For every b, b' in SZ, $B = \langle b, b', Z\rangle$ has finite index over its centre. So by Schur's theorem (§9.2), B' is finite and hence B' = 1.

<u>THEOREM 6. (Swan). If G is finitely presentable, non-abelian and cd G $\leq$ 2, then cd $\zeta_1(G) < 2$ (i.e., $\zeta_1(G)$ is trivial or infinite cyclic).</u>

This result applies to all finitely generated one-relator groups and to all knot groups.

The proof depends on the following proposition which we shall prove in the next section.

<u>PROPOSITION 13. Assume $0 \to P_k \to \dots \to P_o \to \mathbb{Z} \to 0$ is a G-projective resolution with all P_i finitely generated. (Then $H_i(G,\mathbb{Z})$ is finitely generated for each i.) If $h_i = h(H_i(G,\mathbb{Z}))$ (= the torsion-free rank) and $\Sigma(-1)^i h_i \neq 0$, then cd $\zeta_1(G) \leq$ cd G - 2 or $\zeta_1(G) = 1$.</u>

<u>Problem.</u> Can the inequality of Proposition 13 be sharpened?

<u>Exercise.</u> (Murasugi [18]. Cf. also [19].) Prove that any torsion-free one-relator group G has trivial centre if its minimum number of generators is more than 2. (Use Proposition 13. Cf. Lemma 16 §8.11.) We remark that every one-relator group with elements of finite order has trivial centre: cf. [11], Theorem 4.

LEMMA 9. <u>Assume cd $G = $ cd $\zeta_1(G) = k$ and G/G' is non-periodic. Then there exists a subgroup H containing G' such that cd $H = $ cd $\zeta_1(H) = k-1$.</u>

<u>Proof.</u> By Proposition 12, G/Z is periodic ($Z = \zeta_1(G)$). Hence $G'Z/G'$ is not periodic and has finite torsion-free rank (because it is isomorphic to an image of Z). Hence we can find $G' \leq H < K \leq G'Z$ so that $G'Z/K$ is periodic and K/H is infinite cyclic. Then

$$K \cap Z = A \times (H \cap Z),$$

where A is infinite cyclic. Now

$$\begin{aligned} \text{cd } Z &= \text{cd } K \cap Z, \text{ because } Z/K \cap Z \text{ is periodic,} \\ &= 1 + \text{cd}(H \cap Z), \text{ by Lemma 6, p.152.} \end{aligned}$$

Therefore cd $H \geq$ cd $\zeta_1(H) \geq k-1$. But $\langle A,H \rangle = A \times H$ and cd $(A\times H) = 1 + $ cd $H \leq k$, i.e. **cd $H \leq k-1$**. Consequently H has the required properties.

LEMMA 10. If cd G = 1 and G is not cyclic, then $\zeta_1(G) = 1$.

Of course this is implied by Stallings' theorem [27]. But we may argue simply and directly as follows.

Proof. (Swan) Assume $\zeta_1(G) \neq 1$. By Proposition 12(i), every non-trivial subgroup has a non-trivial centre and so it will be sufficient to assume G is finitely generated. By Lemma 9 we can restrict attention to G/G' finite. But then $h_1 = 0$ and Proposition 13 applies because the augmentation ideal $\mathfrak{g}$ is projective and finitely generated.

Proof of Theorem 6. Take a finite presentation

$$1 \to R \to F \to G \to 1.$$

Then $$0 \to \mathfrak{r}/\mathfrak{f}\mathfrak{r} \to \mathfrak{f}/\mathfrak{f}\mathfrak{r} \to ZG \to Z \to 0$$

is a projective resolution with all terms finitely generated (cf. Remark (7), p.37). If G/G' is finite,

$$1 - h_1 + h_2 = 1 + h_2 \neq 0$$

and so, by Proposition 13, $\zeta_1(G) = 1$.

Assume therefore that G/G' is not finite and cd $\zeta_1(G) = 2$. Then Lemma 9 produces a subgroup H which must be cyclic by Lemma 10. But $H \geqslant G'$ and so G is soluble but not abelian by hypothesis. This contradicts Proposition 12(ii).

As another application of Lemma 9 we establish the inequality (*) for groups in the class $\mathfrak{P}$ as defined on p.61.

PROPOSITION 14. If G is a non-abelian group in $\mathfrak{P}$, then cd G > cd $\zeta_1(G)$.

Proof. Let $Z = \zeta_1(G)$ and assume cd G = cd Z = k. Choose $Z_0 \leq Z$ to be a finitely generated subgroup of the same rank as Z. Then choose a non-abelian finitely generated subgroup M containing Z_0. Since $G \in \mathfrak{P}$, there exists $N \triangleleft M$ so that M/N is infinite cyclic, say on xN. As G/Z is periodic, $x^s N \in ZN/N$ for some finite s. Hence $ZN/(ZN)'$ is non-periodic and so, by Lemma 9, we have a subgroup $H \geqslant (ZN)'$ with cd H = cd $\zeta_1(H)$ = k-1. Since $H \in \mathfrak{P}$, by an induction on k we may assume H is abelian. But then ZM is soluble and so is abelian (Proposition 12(ii)), which contradicts our choice of M.

8.10 Euler characteristics.

Let R be a ring. We shall have to assume that R satisfies the invariance property (ip): any two bases of a finitely generated free R-module have the same number of elements. Obviously, if a homomorphic image of R has ip then so does R. Now all fields have ip and so all commutative rings have. Hence so do all group rings. In our applications R will be $\mathbb{Z}G$ for some group G.

Denote by C the additive subgroup of R generated by all

$xy - yx$ and by $x \mapsto \bar{x}$ the natural additive homomorphism $R \to R/C$.

If $(a_{ij}) \in R^{n\times n}$ (matrices), define the <u>trace</u> by

$$\mathrm{tr}(a_{ij}) = \sum_{i=1}^{n} \bar{a}_{ii}.$$

Then

(Tr 1) tr is an additive homomorphism: $R^{n\times n} \to R/C$ (obvious);

(Tr 2) if $a \in R^{p\times q}$, $b \in R^{q\times p}$, $\mathrm{tr}(ab) = \mathrm{tr}(ba)$ (easy).

If M is a finitely generated free R-module and $\varphi \in \mathrm{End}_R M$, we can define a trace of φ thus: choose any R-basis $v_1,\dots,v_n$ of M and let

$$a = (\varphi;\ (v_i)) = \text{matrix of } \varphi \text{ with respect to } (v_i),\ (v_i).$$

Suppose $w_1,\dots,w_n$ is any other basis of M and $(\varphi;\ (w_i)) = b$. (Note that by ip we have the same number of w's as v's.) Then

$$b = (1_M;\ (v_i)(w_i))\ a\ (1_M;\ (w_i)(v_i)) = u^{-1} a\ u\ ,$$

where $u = (1_M;\ (w_i)(v_i))$ = matrix of identity with respect to $(w_i),(v_i)$. Now

$$\begin{aligned} \mathrm{tr}\ b &= \mathrm{tr}(u^{-1} a\ u) \\ &= \mathrm{tr}\ (auu^{-1}) \quad \text{by (Tr 2)} \\ &= \mathrm{tr}\ a. \end{aligned}$$

We can therefore unambiguously define tr φ to be tr a. This function also satisfies (Tr 1), (Tr 2).

Next let P be a finitely generated projective R-module and

$\varphi \in \mathrm{End}_R P$. We shall define a trace of φ. Let $P \oplus Q = M$, with M free. Extend φ to M by $(p,q)\varphi' = (p\varphi, 0)$; i.e., $\varphi' = \varphi + 0_Q$. We show tr φ' is independent of Q. Suppose $P \oplus Q' = M'$, with M' also free. Then

$$Q \oplus M' \cong Q \oplus P \oplus Q' \cong Q' \oplus M.$$

Let α be any one isomorphism: $Q \oplus M' \xrightarrow{\sim} Q' \oplus M$. Then

$$1_P + \alpha : \ P \oplus Q \oplus M' \xrightarrow{\sim} P \oplus Q' \oplus M.$$

Now $\mathrm{tr}(\varphi + 0_Q) = \mathrm{tr}(\varphi + 0_Q + 0_{M'})$ and $\mathrm{tr}(\varphi + 0_{Q'}) = \mathrm{tr}(\varphi + 0_{Q'} + 0_M)$. But

$$\varphi + 0_Q + 0_{M'} = (1_P + \alpha)(\varphi + 0_{Q'} + 0_M)(1_P + \alpha)^{-1}$$

and so $\mathrm{tr}(\varphi + 0_{Q'}) = \mathrm{tr}(\varphi + 0_Q)$. We call this common value tr φ, the trace of φ. This still satisfies (Tr 1), (Tr 2).

Special case: We write $\mathrm{tr}(1_P) = \chi(P)$ and call this the <u>Euler characteristic of P</u>.

(Note: When P is free, this is precisely (rank P)1 + C. Observe that this could be zero: e.g. if R is a field of characteristic p and P is a vector space of dimension p.)

A <u>finite resolution</u> of an R-module A shall mean a resolution of finite length:

$$P.: \ 0 \to P_n \to \dots \to P_o \to A \to 0$$

where each P_i is finitely generated over R. (Hence A is necessarily finitely generated.)

If P. is a finite projective resolution we define the Euler

characteristic $\chi(P.)$ of P. to be

$$\chi(.) = \sum_{i=0}^{n} (-1)^i \chi(P_i).$$

PROPOSITION 15. *If P., P'. are finite projective resolutions of A, then $\chi(P.) = \chi(P'.)$.*

Definition. We call the common value given by Proposition 15 *the Euler characteristic of A.* (Note that when A is projective this notation and terminology reduces to the previous.)

LEMMA 11. *(Schanuel Lemma.) Given two short exact sequences of G-groups (cf. p.1):*

$$1 \to K \to P \xrightarrow{\pi} A \to 1$$
$$1 \to K_1 \to P_1 \xrightarrow{\pi_1} A \to 1.$$

If π lifts to $\rho: P \to P_1$ and π_1 lifts to $\rho_1: P_1 \to P$, then

$$P[K_1 \cong P_1[K \quad \text{(as G-groups)}.$$

Proof. If

$$\begin{array}{ccc} E & \longrightarrow & P \\ \downarrow & & \downarrow \pi \\ P_1 & \xrightarrow{\pi_1} & A \end{array}$$

is a pull-back square, then E splits over K_1 by the existence of ρ and also over K by the existence of ρ_1.

Remark. In our present application of this lemma the G-groups are G-modules and P, P_1 are G-projective. The existence of ρ, ρ_1 is then assured and the conclusion reads $P \oplus K_1 \cong P_1 \oplus K$.

LEMMA 12. Given the exact sequences of G-modules

$$0 \to K \to P_n \to \cdots \to P_1 \to P_o \to A \to 0$$

$$0 \to L \to Q_n \to \cdots \to Q_1 \to Q_o \to A \to 0$$

where P_i, Q_i are projective, then

$$K \oplus Q_n \oplus P_{n-1} \oplus \cdots \cong L \oplus P_n \oplus Q_{n-1} \oplus \cdots \; .$$

Proof. Let

$$\begin{array}{ccccccc} P_1 & & \longrightarrow & & P_o & \to A \to 0, \\ & \searrow & & \nearrow & & \\ & & \bar{K} & & & \\ & \nearrow & & \searrow & & \\ 0 & & & & 0 & \end{array}$$

and similarly define $\bar{L}$.

By Lemma 11, $\bar{K} \oplus Q_o \cong \bar{L} \oplus P_o$. Now

$$0 \to K \to P_n \to \cdots \to P_2 \to P_1 \oplus Q_o \to \bar{K} \oplus Q_o \to 0$$

$$0 \to L \to Q_n \to \cdots \to Q_2 \to Q_1 \oplus P_o \to \bar{L} \oplus P_o \to 0.$$

So induction gives the result.

Proof of Proposition 15. Let P. and P'. be as given in Lemma 12 but with $K = L = 0$. (We do not a priori assume P. and P'. have the same length: if not, fill up with zeros.) Then

$$Q_n \oplus P_{n-1} \oplus \cdots \cong P_n \oplus Q_{n-1} \oplus \cdots \; .$$

Therefore

$$\mathrm{tr}(1_{Q_n} + 1_{P_{n-1}} + \cdots) = \mathrm{tr}(1_{P_n} + 1_{Q_{n-1}} + \cdots) \; ,$$

i.e. $$\chi(Q_n) + \chi(P_{n-1}) + \cdots = \chi(P_n) + \chi(Q_{n-1}) + \cdots \; .$$

We can view R/C naturally as a module over the centre of R: given $x \in R$, $z \in$ centre R, let $\bar{x}.z = \overline{xz}$. This is unambiguous because $(xy-yx)z = x(yz) - (yz)x \in C$, all $x,y \in R$.

<u>THEOREM 7. (Stallings[25] (3.4, p.133); Kaplansky [10] (4.7, p.33).) If R has ip and A has a finite projective resolution, then for any $y \in (\text{ann } A) \cap (\text{centre } R)$, $\chi(A).y = 0$.</u>

<u>LEMMA 13. Given</u>

$$\begin{array}{ccccccccc} \cdots \to & P_2 & \xrightarrow{f_2} & P_1 & \xrightarrow{f_1} & P_o & \xrightarrow{f_o} & A & \longrightarrow 0 \\ & {\scriptstyle h_2}\downarrow & & {\scriptstyle h_1}\downarrow & & {\scriptstyle h_o}\downarrow & & {\scriptstyle 0}\downarrow & \\ \cdots \to & Q_2 & \xrightarrow{g_2} & Q_1 & \xrightarrow{g_1} & Q_o & \xrightarrow{g_o} & A & \longrightarrow 0 \end{array}$$

<u>with both rows exact and all P_i projective, then there exist R-homomorphisms $s_o = 0, s_1, s_2, \ldots, s_i : P_{i-1} \to Q_i$ (homotopies) such that $h_i = s_{i+1}g_{i+1} + f_i g_i$, $i \geqslant 0$.</u>

<u>Proof</u>. There exists s_1 (by the projectivity of P_o) so that

$$h_o = s_1 g_1 + f_o s_o ;$$

then there exists s_2 (by the projectivity of P_1 and using $h_1 - f_1 s_1 : P_1 \to Q_1$) so that

$$h_1 = s_2 g_2 + f_1 s_1 ; \quad \text{etc..}$$

<u>Proof of Theorem 7</u>. Write down the finite resolution twice:

$$\begin{array}{ccccccc} 0 \to & P_n & \to \cdots \to & P_o & \to & A & \to 0 \\ & {\scriptstyle y}\downarrow & & {\scriptstyle y}\downarrow & & {\scriptstyle y}\downarrow & \\ 0 \to & P_n & \to \cdots \to & P_o & \to & A & \to 0 . \end{array}$$

The down maps are multiplication by y: this is a module

endomorphism as $y \in$ centre R; and it is zero on A as $y \in$ ann A.

Construct homotopies s_i as in Lemma 13. Here $f_i = g_i$ and so $tr(s_{i+1}f_{i+1} + f_i s_i) = tr(s_{i+1}f_{i+1}) + tr(f_i s_i)$ by (Tr 1)

and $tr(s_{i+1}f_{i+1}) = tr(f_{i+1}s_{i+1})$ by (Tr 2).

Thus

$$0 = \sum_{i=0}^{n} (-1)^i tr(s_{i+1}f_{i+1} + f_i s_i) = \Sigma(-1)^i tr(1_{P_i} y) ;$$

But $tr(1_{P_i} y) = \chi(P_i)y$ and so the right hand side is $\chi(A)y$.

<u>We now assume that R = $\mathbb{Z}G$ for some group G.</u> If $a \in \mathbb{Z}G$, let S(a) be the set of all elements of G occuring in a. Denote the conjugacy class containing x by $\tilde{x}$. Then C is all elements a with coefficient sum zero on every $S(a) \cap \tilde{x}$: for if $a, b \in G$, $ab - ba = ab - (ab)^a$; and conversely, $a^x - a = x^{-1}(ax) - (ax)x^{-1}$. Thus R/C has $\mathbb{Z}$-basis $\bar{a}_1, \bar{a}_2, \dots$, where $a_1, a_2, \dots$ are representatives of all the distinct conjugacy classes of G. We shall usually identify R/C with the $\mathbb{Z}$-module freely generated by the conjugacy classes $\tilde{x}$ of G.

Further, the centre of $\mathbb{Z}G$ consists of all elements in $\mathbb{Z}G$ whose coefficients for each conjugacy class are constant.

Thus (centre $\mathbb{Z}G$) $\cap$ C = 0.

<u>COROLLARY (to Theorem 7). If the $\mathbb{Z}G$-module A has a finite FREE resolution and $\chi(A) \neq 0$, then (ann A) $\cap$ (centre $\mathbb{Z}G$) = 0. In particular, if $\mathbb{Z}$ has a finite FREE resolution and $\chi(\mathbb{Z}) \neq 0$,</u>

<u>then $\zeta_1(G) = 1$.</u>

<u>Proof</u>. Here $\chi(A) = m1 + C$ for some integer m. So $\chi(A)y = 0$ implies $my \in C$. But $my \in$ centre $\mathbb{Z}G$ and so $my = 0$. Therefore $y = 0$.

<u>PROPOSITION 16. (Swan.) If $\mathbb{Z}$ has a finite projective resolution and $\chi(\mathbb{Z}) \neq 0$, then $\operatorname{cd} \zeta_1(G) \leq \operatorname{cd} G - 2$ or $\zeta_1(G) = 1$.</u>

<u>Proof</u>. As $\chi(\mathbb{Z}) \neq 0$, there exists a conjugacy class $\tilde{x}$ whose coefficient in $\chi(\mathbb{Z})$ is not 0. Now $\chi(\mathbb{Z})z = \chi(\mathbb{Z})$ for each z in $\zeta_1(G)$ (Theorem 7). Hence the coefficient of $\widetilde{xz^k}$ is also non-zero, for all k. But $\chi(\mathbb{Z})$ can only involve a finite number of conjugacy classes and so there exists m such that $\tilde{x} = \widetilde{xz^m}$. Hence $g^{-1}xg = xz^m$ for some g.

Let $z_1,\ldots,z_r$ be a maximal $\mathbb{Z}$-independent set in $Z = \zeta_1(G)$. For each i, find g_i so that $g_i^{-1}xg_i = xz_i^{m_i}$ for some m_i. Let $H = \langle Z, x, g_1\rangle$. No power of x is in Z, because

$$[x^k, g_1] = [x, g_1]^k = z_1^{m_1 k} ;$$

and for a similar reason, no power of g_1 is in $\langle Z, x\rangle$. Hence $\langle z_1,\ldots,z_r, x, g_1\rangle$ is polycyclic of Hirsch number r+2. Since H is nilpotent, we conclude $\operatorname{cd} H = \operatorname{cd} Z + 2$, by Theorem 5.

Proposition 16 leads directly to Proposition 13 of the last section provided we can show $\Sigma(-1)^i h_i \neq 0$ implies $\chi(\mathbb{Z}) \neq 0$. This is an immediate consequence of Lemma 15 below.

<u>Definition</u>. Let P. be a projective resolution of $\mathbb{Z}$ and assume P_i is finitely generated for $i = 0,1,\ldots,n$. Define the <u>n-th partial Euler characteristic</u> $\chi_n(P.)$ to be

$$\chi_n(P.) = \sum_{i=0}^{n} (-1)^i \chi(P_i).$$

Notational reminders: (1) For any abelian group A, h(A) is the torsion-free rank (= Q-dimension of $A \otimes Q$); (2) $h_i = h(H_i(G,\mathbb{Z}))$; (3) for any G-module M, $M_G = M/M\mathfrak{g} = M \otimes_G \mathbb{Z}$.

We also define $\bar{\epsilon}: \mathbb{Z}G/C \longrightarrow \mathbb{Z}$ to be the homomorphism induced by the augmentation $\epsilon: \mathbb{Z}G \longrightarrow \mathbb{Z}$ $(C \leq \mathfrak{g})$.

<u>LEMMA 14. If P is a finitely generated projective module, then</u>

$$\chi(P)\bar{\epsilon} = h(P/P\mathfrak{g}).$$

<u>Proof</u>. Let $P \oplus Q = M$, free. Then $P/P\mathfrak{g}$ is $\mathbb{Z}$-free. Let (v_i) be a $\mathbb{Z}G$-basis of M and $(1_P + 0_Q; (v_i)) = (a_{ij})$. Then

$$\chi(P)\bar{\epsilon} = \Sigma\bar{a}_{ii}\bar{\epsilon} = \Sigma a_{ii}\epsilon = \mathrm{tr}(a_{ij}\epsilon).$$

<u>LEMMA 15. (Swan [31], §1.) If P. is a projective resolution of $\mathbb{Z}$ and $\chi_n(P.)$ is defined, then</u>

$$\chi_n(P.)\bar{\epsilon} = \sum_{i=0}^{n-1} (-1)^i h_i + (-1)^n h(\mathrm{Ker}(P_{n,G} \to P_{n-1,G})).$$

<u>Proof</u>. Let $Q_i = \mathrm{Im}(P_{i+1} \longrightarrow P_i)$. Then

$$0 \longrightarrow H_{i+1}(G,\mathbb{Z}) \longrightarrow Q_{i,G} \longrightarrow P_{i,G} \longrightarrow Q_{i-1,G} \longrightarrow 0$$

is exact for all $i \geqslant 0$. (We interpret $Q_{-1,G} = \mathbb{Z}$.) Therefore

$$h_{i+1} - h(Q_{i,G}) + h(P_{i,G}) - h(Q_{i-1,G}) = 0.$$

Multiply by $(-1)^i$ and add for $i = 0,1,\ldots,n-1$:

$$- h_0 + h_1 + \ldots + (-1)^{n-1}h_n + (-1)^n h(Q_{n-1,G}) + \chi_{n-1}(P.)\bar{\varepsilon} = 0.$$

But if $K = \mathrm{Ker}\,(P_{n,G} \to P_{n-1,G})$,

$$h_n - h(Q_{n-1,G}) = - h(P_{n,G}) + h(K)$$

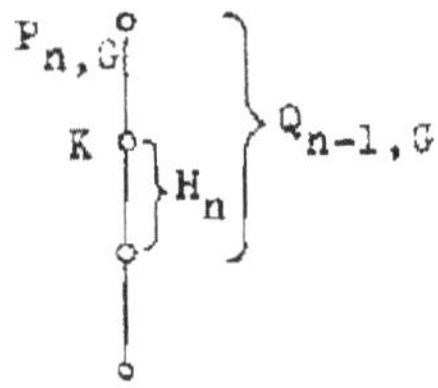

and hence

$$\chi_n(P.)\bar{\varepsilon} = (-1)^n h(K) + \sum_{i=0}^{n-1} (-1)^i h_i,$$

as required.

8.11 Trivial cohomological dimension.

Definition. G has trivial cohomological dimension k (we write tcd G = k) if $H^q(G,A) = 0$ for all $q > k$ and all trivial G-modules A and there exists a trivial A such that $H^k(G,A) \neq 0$.

The Universal Coefficient Theorem (p.49) immediately implies

PROPOSITION 17. tcd $G \leqslant k$ if, and only if, $H_q(G,\mathbb{Z}) = 0$ all $q > k$, $H_k(G,\mathbb{Z})$ is $\mathbb{Z}$-free.

There do exist groups of trivial cohomological dimension equal to any given positive integer. For by the corollary to Lemma 6, p. 152 if G is finitely generated, torsion-free nilpotent, then tcd G = cd G. We shall see later (corollary to Theorem 8) that there exist $G \neq 1$ with tcd $G = 0$.

<u>No finite group can have finite tcd</u>. This follows from a theorem of Swan [30] that if G is finite, then $H_k(G,\mathbb{Z}) \neq 0$ for infinitely many k.

<u>**Warnings**</u>. (i) tcd $G \leq k$ is not equivalent to $H^{k+1}(G,A) = 0$ for all trivial A. E.g. any finite perfect G with $H_2(G,\mathbb{Z}) \neq 0$. (Explicit example: G is the alternating group of degree 5. Here $|H_2(G,\mathbb{Z})| = 2$: Schur [22].)

(ii) $H \leq G$ need not imply tcd $H \leq$ tcd G. Example: If $G = \langle a,b \mid a^3 = b^2 \rangle$, then cd $G \leq 2$ (Theorem 2, §8.4 or Theorem 4, §8.6) and $H = \langle b^2, ab \rangle$ is free abelian of rank 2. So cd H = tcd H = 2 and cd G = 2. But tcd G = 1 by Proposition 18 below.

<u>LEMMA 16. Let $1 \to R \to F \to G \to 1$ be a free presentation with $R = \langle F\langle w \rangle$ (= normal closure of w). Then $H_2(G,\mathbb{Z}) = 0$ if, and only if, $R \not\leq F'$.</u>

<u>Proof</u>. By Magnus (Theorem 2, p.57), $R \neq [R,F]$ and here $R/[R,F]$ is cyclic. Now $R/R \cap F' \cong RF'/F'$ is infinite cyclic, but possibly trivial. Therefore $R \neq R \cap F'$ if, and only if, $R \cap F' = [R,F]$, i.e., $w \notin F'$ if, and only if, $H_2(G,\mathbb{Z}) = 0$.

PROPOSITION 18. ([3], Lemma 4.) Let G be a one-relator group. Then the following two conditions are equivalent:

(i) cd $G \le 2$ and tcd $G = 1$;

(ii) G/G' is non-trivial free abelian and the relation cannot be expressed as a word in commutators.

Proof. Assume a presentation as in Lemma 16.

(i) ⇒ (ii). tcd $G = 1$ implies (a) $H_2(G,\mathbb{Z}) = 0$ and (b) $H_1(G,\mathbb{Z})$ is non-trivial free abelian. So (b) says G/G' is non-trivial free abelian and (a) implies $w \notin F'$ by Lemma 16.

(ii) ⇒ (i). We assert G is torsion-free: if $w = s^n$, then as $R \not\leq F'$, $w \notin F'$ and $\bar{w} = wF'$ generates a non-trivial infinite cyclic subgroup in F/F'. But $s^n \in RF'$ and $F/RF' \cong G/G'$ is torsion-free. Therefore $s \in RF'$ and so $\langle \bar{s} \rangle = \langle \bar{w} \rangle$, whence $n = \pm 1$.

Thus cd $G \le 2$ by Lyndon (p.129). Also $H_2(G,\mathbb{Z}) = 0$ by Lemma 16 so that tcd $G = 1$ (Proposition 17).

Exercises.

(1) If G is nilpotent and tcd $H = 1$ for all subgroups H then G is infinite cyclic. (Hint: look at the abelian subgroups.)

(2) ([3], Lemma 1, p.402.) If cd $G = k$, $M \triangleleft G$ and tcd $M = 0$, then cd $G/M \le k$. (Hint: Use the inf-res sequence, p.93).

Generalisations of exercise 1 can be found in [26], [28], [3].

We turn now to groups of tcd = 0. First some preliminary general facts. We recall that for any G, d(G) stands for the minimum number of generators of G; and $d_P(G)$ is the minimum number of P-generators of the P-group G (cf. p.98). (Note that for abelian A, $h(A) = d_{\mathbb{Q}}(A \otimes \mathbb{Q})$.)

If G has a finite presentation with s generators and r relations, we call s-r the <u>deficiency of the presentation</u> and we define the <u>deficiency of G</u>:

def G = sup {deficiency of Pr; all finite presentations Pr}.

Note that def G is defined whenever G has a finite presentation. It is not immediately obvious that def G is then necessarily finite. However, this is true, as we shall see in the corollary to Lemma 17.

If P. is a projective resolution of $\mathbb{Z}$ and $\chi_n(P.)$ is defined (cf. end of last section), we shall set

$$\mu_n(P.) = (-1)^n \chi_n(P.)\bar{\varepsilon}.$$

Then Lemma 15 reads

$$\mu_n(P.) = \sum_{i=0}^{n-1} (-1)^{n-i} h_i + h(K) . \qquad (*)$$

Define

$$\mu_n(G) = \inf \{\mu_n(P.); \text{ all possible } P.\}.$$

This exists provided $\mu_n(P.)$ exists for some P.

LEMMA 17. If $\mu_n(G)$ is defined, there exists P. so that $\mu_n(G) = \mu_n(P.)$.

Proof. In (*) we have $K = \mathrm{Ker}(P_{n,G} \to P_{n-1,G})$ and so is $\mathbb{Z}$-free. Hence $h(K) = d(K) \geq d(H_n(G,\mathbb{Z})) = d_n$ and therefore

$$\mu_n(G) \geq d_n + \sum_{0}^{n-1} (-1)^{n-i} h_i \; . \qquad (**)$$

COROLLARY. If def G is defined, there exists a presentation Pr so that def G = def Pr.

Proof. From any finite presentation

$$\mathrm{Pr:}\;\; 1 \to R \to F \to G \to 1$$

we obtain

$$M \to \mathfrak{f}/\mathfrak{f}\mathfrak{r} \to \mathbb{Z}G \to \mathbb{Z} \to 0,$$

where M is G-free on a set in one-one correspondence with the relations (cf. p.37). If this exact sequence is continued to a projective resolution P.,

$$\mu_2(P.) = 1-s+r \geq \mu_2(G).$$

So

$$\begin{aligned} \mathrm{def}\, G &\leq 1 - \mu_2(G) \\ &< \infty \quad \text{by the lemma.} \end{aligned}$$

PROPOSITION 19. (Swan [31], §1.) (1) $\mu_1(G) \leq d(G) - 1$;

(2) $\mu_2(G) \leq 1 - \mathrm{def}\, G$;

(3) $\mathrm{def}\, G \leq h(H_1(G,\mathbb{Z})) - d(H_2(G,\mathbb{Z}))$.

__Proof.__ (1) Choose a presentation with free group F so that $d(F) = d(G)$. Then the corresponding resolution $P.$ gives

$$\mu_1(P.) = d(G) - 1.$$

(2) Choose a presentation Pr so that def Pr = def G. The corresponding $P.$ gives $\mu_2(P.) = 1 - \operatorname{def} G$.

(3) $\operatorname{def} G \leq 1 - \mu_2(G)$ by (2)

$\leq 1 - \{d(H_2(G,\mathbb{Z})) - h_1 + 1\}$ by (**) above.

Swan proves in [31] that __if G is finite, $1 + \mu_1(G) = d_G(\mathfrak{g})$.__ This is rather hard. An immediate consequence is that if G is finite and nilpotent, then (1) in Proposition 19 is an equality.

__Exercise.__ (Magnus: cf [29].) If G has a presentation with $r+s$ generators and r relations and G can also be generated by s elements, prove that G is free of rank s. (Hint: Use (3) of Proposition 19 to show that $H_2(G,\mathbb{Z}) = 0$ and G/G' is free abelian of rank s.)

THEOREM 8. __If $\infty > \operatorname{def} G \geq h(G/G')$, then $H_2(T,\mathbb{Z}) = 0$, where T/G' is the torsion group of G/G'.__

__LEMMA 18. (Swan [31], §3.)__ __Let $T \triangleleft G$ be such that $Q = G/T$ is free abelian and assume $\mu_n(G)$ is defined. Then__

$$\underline{\mu_n(G) \geq \delta + \sum_{0}^{n-1} (-1)^{n-i} h_i ,}$$

where δ is the smallest integer so that $H_n(T,\mathbb{Z})$ is an image of a torsion-free $\mathbb{Z}Q$- module of rank δ.

(Q acts on the homology of T in essentially the same way as it does on the cohomology of T. Hence $H_n(T,\mathbb{Z})$ is a Q-module. Further, $\mathbb{Z}Q$ is an integral domain (Theorem 3, p.61), so there is a quotient field F and any Q-module B has a rank $h_Q(B) = d_F(B \otimes F)$ = F-dimension of $B \otimes F$. Thus δ is defined ($\leq \infty$).)

Proof. By Lemma 17 and Lemma 15 (= (*)), there exists P. such that

$$\mu_n(G) = \mu_n(P.) = \Sigma(-1)^{n-i}\, h_i + h(K),$$

where $K = \mathrm{Ker}(P_{n,G} \xrightarrow{\varphi} P_{n-1,G})$.

Now for any G-module A, A_T is a Q-module and $A_G \cong (A_T)_Q$. If B is any Q-module, clearly $h_Q(B) \geq h(B_Q)$. There is equality if B is Q-free and hence there is also equality if B is Q-projective.

Let $L = \mathrm{Ker}(P_{n,T} \xrightarrow{\psi} P_{n-1,T})$. Then the commutativity of

$$\begin{array}{ccc} P_{n,T} & \xrightarrow{\psi} & P_{n-1,T} \\ \downarrow & & \downarrow \\ P_{n,G} & \xrightarrow{\varphi} & P_{n-1,G} \end{array}$$

gives $h_Q(\mathrm{Im}\,\psi) \geq h(\mathrm{Im}\,\varphi)$. Hence

$$\begin{aligned} h(K) &= h(P_{n,G}) - h(\mathrm{Im}\,\varphi) \\ &\geq h(P_{n,G}) - h_Q(\mathrm{Im}\,\psi) \\ &= h_Q(P_{n,T}) - h_Q(\mathrm{Im}\,\psi), \text{ since } P_{n,T} \text{ is Q-projective,} \\ &= h_Q(L) \\ &\geq \delta, \quad \text{by the definition of } \delta. \end{aligned}$$

Proof of Theorem 8. Let $k = h(G/G') = h_1$. Then

$$1 - \text{def } G \geq \mu_2(G) \quad \text{(Proposition 19)}$$
$$\geq \delta - k + 1,$$

by Lemma 18. But def $G \geq k$ and so $\delta = 0$. Thus $H_2(T,\mathbb{Z}) = 0$.

COROLLARY. If cd $G \leq 2$, G' is perfect and G/G' is free abelian of rank $\leq$ def G, then tcd $G' = 0$.

This corollary covers all knot groups G with perfect commutator group ([31], §3): for such groups have def $G = h(G/G') = 1$ [7].

An explicit one-relator group as in the corollary is

$$G = \langle a,b \mid [[a,b],[a^2,b]] = a \rangle.$$

(Cf. [3] Theorem 4 and Example 2.)

8.12 Finite groups.

Definition. Let $\mathcal{C}$ be a category of G-modules. We define the $\mathcal{C}$-cohomological dimension of G, $\mathcal{C}$-cd G, to be the smallest integer k so that $H^q(G,A) = 0$ for all $q > k$ and all A in $\mathcal{C}$.

If $\mathcal{C} = \text{Mod}_G$, then $\mathcal{C}$-cd = cd; and if $\mathcal{C}$ = all trivial G-modules, then $\mathcal{C}$-cd = tcd.

Suppose K is a commutative ring and $\mathcal{C}$ is all KG-modules and KG-homomorphisms. We shall write $\mathfrak{C}$-cd = K-cd.

The Shapiro Lemma (p.92) shows that K-cd H $\leq$ K-cd G for all subgroups H of G: for if A is a KH-module, then

$$A\uparrow_H^G = \mathrm{Hom}_{\mathbb{Z}H}(\mathbb{Z}G, A)$$

is a KG-module. If K-cd G is finite and H is cyclic, then the explicit formulae on p.40 show that multiplication by $|H|$ is an automorphism of K. Thus we have

PROPOSITION 20. <u>If K-cd G is finite, then K is uniquely divisible by the order of every finite cyclic subgroup of G.</u>

PROPOSITION 21. <u>The following are equivalent:</u>

<u>(i) K-cd G = 0;</u>

<u>(ii) G is finite and K-cd G is finite;</u>

<u>(iii) G is finite and $|G| = n$ is a unit in K.</u>

<u>Proof.</u> <u>(i) ⇒ (ii).</u> If K-cd G = 0, then K is KG-projective and so $KG \xrightarrow{\epsilon} K \longrightarrow 0$ splits: say $\alpha : K \longrightarrow KG$, where $\alpha\epsilon = \mathrm{id}_K$. Now 1α is G-invariant and non-zero. But the only G-invariant elements in KG are the K-multiplies of $\sum_{x \in G} x$. Hence G must be finite.

<u>(ii) ⇒ (iii).</u> By Proposition 20, every prime in n is a unit in K. Hence n is a unit in K.

<u>(iii) ⇒ (i).</u> For any KG-module A and all q > 0, $nH^q(G,A) = 0$ (Corollary 1, p.91). But multiplication by n is a KG-automorphism by hypothesis and hence is also an automorphism

of $H^q(G,A)$.   Thus $H^q(G,A) = 0$.

Exercises.

(1)   If F is a field of characteristic zero and G is finite, prove
that for all $q > 0$, $H^q(G, F/\mathbb{Z}) \simeq H^{q+1}(G,\mathbb{Z})$.

(2)   If $H^q(G,A) = 0$ for all $q > k$ and all torsion G-modules A,
then G is torsion-free.

(3)   If G is finite and $H^q(G,A) = 0$ for all $q > k$ and all finite
G-modules A, then $G = 1$.

Exercise 3 shows that if Tor is the full subcategory of all
torsion G-modules and G is finite, then Tor-cd G cannot be finite.
Nevertheless, it is true and of considerable importance that
inverse limits of finite groups can indeed have finite cohomological
dimension relative to Tor provided the cohomology is interpreted
in the correct way.   We explain this in brief outline.

Suppose (U_i) is the family of all normal subgroups of
finite index in a group G.   Let Dis be the category of all
G-modules A so that $A = \cup A^{U_i}$.   Thus $\cap U_i$ lies in the kernel of
the representation of G on A.   The family of representations
$G/U_i \longrightarrow \text{Aut } A^{U_i}$ is an inverse system yielding a profinite
representation of $\widehat{G} = \varprojlim G/U_i$ on A.   The category Dis can
therefore be viewed as a category of $\widehat{G}$-modules: these are the
discrete G-modules.

The profinite cohomology groups are defined by the formula

$$\mathrm{pr}H^q(\hat{G},A) = \varinjlim H^q(G/U_i,\ A^{U_i}),$$

using the obvious inflations for the direct limit (p.89). There is a good cohomology theory of profinite groups and much is known about cohomological dimension. If $\mathcal{J}$ = Tor $\cap$ Dis, then the cohomological dimension of G is defined with respect to $\mathcal{J}$. It would be of interest to explore its relation with $\mathcal{J}$-cd G. (The best reference for all this is Serre, Cohomologie galoisienne, Springer Lecture Notes 5.)

<u>PROPOSITION 22. Let G be finite and K a field of characteristic dividing $|G|$. If A is a finitely generated KG-module not in the first block of KG, then $H^q(G,A) = 0$ for all $q \geqslant 0$.</u>

<u>Explanation</u>. The first block of KG is the indecomposable two-sided ideal having K as a module image. Our hypothesis on A says that no composition factor of A is a composition factor of the first block. (For the elementary theory of blocks read [8], chapter 8.)

<u>Preliminaries</u>. Let J be the Jacobson radical of KG. If M is a finitely generated KG-module, there exists a finitely generated projective KG-module P so that φ: $P/PJ \xrightarrow{\sim} M/MJ$. By projectivity, we can lift φ to ψ: $P \longrightarrow M$ and ψ is epimorphic by the corollary on p.98. Moreover, Ker $\psi \leq PJ$. We call ψ a <u>minimal epimorphism</u>.

Since P is finitely generated, so is Ker ψ and hence we can find a minimal epimorphism $P' \longrightarrow$ Ker ψ. In this way we produce a KG-projective resolution of M in which all maps are minimal. We call this a minimal resolution. Every composition factor occurring in this resolution must lie in the same block as one of the composition factors of M/MJ.

Proof of Proposition 22. We proceed by induction on the composition length of A. If $0 \to A' \to A \to A'' \to 0$ is exact with A" irreducible, then the exact cohomology sequence shows that the result for A' and A" implies it for A.

Suppose therefore that A is irreducible. Take a minimal projective resolution P. of K and let $Y = \mathrm{Im}(P_q \to P_{q-1})$. Then $H^q(G,A)$ is an image of $\mathrm{Hom}_{KG}(Y,A)$. But since Y belongs to the first block and A does not, $\mathrm{Hom}_{KG}(Y,A) = 0$. Thus $H^q(G,A) = 0$ for $q > 0$. Finally, A^G is 0 or A and since A is not K, $A^G = 0$.

Sources and References.

8.2 The results in this section seem to be new. Theorem 1 was suggested by the profinite analogue in [23] and the special case cd $G \leq 1$ observed by Graham Higman in his paper in Oxford Quart. J. Math. 6 (1955), 250-254.

8.3 The proof of the freeness of the resolution (1) in Proposition 4 is essentially that of Balcerzyk's Lemma 3 in [1].

8.5 Swan showed me Proposition 6 in June 1967.

8.6 Barr showed me Theorem 4 in July 1968. The Barr-Beck proof is via the derived functors of Der and is sketched in exercises 1 - 4 at the end of the section.

8.8 Theorem 5 seems to be new. The proof was much simplified by a suggestion of Urs Stammbach.

8.9 The whole of this section owes much to material shown me by Swan in the summer of 1967. In particular, Theorem 6 is his and so are Proposition 13 and Lemma 9 in so far as they are needed for Theorem 6 (i.e., for the case cd $G = 2$).

8.10 The material up to and including Theorem 7 is based on Stalling's work [25]. In my treatment I have also borrowed from Kaplansky [10]. Proposition 16 for the case cd $G = 2$ is an unpublished result of Swan (July 1967): note that this case does not need an appeal to Theorem 5. The partial Euler characteristics were introduced (for free resolutions) by Swan in [31].

8.11 Trivial cohomological dimension seems to make its first appearance in the literature in [3]. Exercise 1 (after Proposition 18) has many ramifications. Stammbach's well-written paper [28] leads to an interesting application in [29]. The treatments of $\mu_n(G)$ and def G are based on Swan [31].

8.12 Proposition 22 seems to be folk-lore.

[1] Balcerzyk, S.: The global dimension of the group rings of abelian groups, Fund. Math. 55 (1964) 293-301.

[2] Barr, M. and Beck, J.: Homology and standard constructions, Springer Lecture Notes, No.80 (1969).

[3] Baumslag, G. and Gruenberg, K.W.: Some reflections on cohomological dimension and freeness, J. of Algebra 6 (1967) 394-409.

[4] Berstein, I.: On the dimension of modules and algebras IX. Direct Limits, Nagoya Math. J. 13 (1958) 83-84.

[5] Bourbaki, N.: Algèbre, chapter 2; Hermann (ASI 1236) 1962.

[6] Cohen, D.E. and Lyndon, R.C.: Free bases for normal subgroups of free groups, Trans. Amer. Math. Soc. 108 (1963) 526-537.

[7] Crowell, R.H. and Fox, R.H.: Introduction to knot theory; Ginn and Co., 1963.

[8] Curtis, C. and Reiner, I.: Representation theory of finite groups and associative algebras; Interscience, 1963.

[9] Hirsch, K.A.: On infinite soluble groups - I, Proc. London Math. Soc. 44 (1938) 53-60.

[10] Kaplansky, I.: Commutative rings, Queen Mary College Math. Notes, 1966.

[11] Karrass, A., Magnus, W. and Solitar, D.: Elements of finite order in groups with a single defining relation, Comm. in Pure and Appl. Math. 13 (1960) 57-66.

[12] Kurosh, A.G.: The theory of groups, vol.2; Chelsea, 1960.

[13] Lyndon, R.C.: Cohomology theory of groups with a single defining relation, Annals of Math. 52 (1950) 650-665.

[14] Lyndon, R.C.: Dependence and independence in free groups, Crelle 210 (1962) 148-174.

[15] Magnus, W., Karrass, A. and Solitar, D.: Combinatorial group theory, Interscience, 1966.

[16] Mal'cev, A.I.: On some classes of infinite soluble groups, Mat.Sb. N.S. 28 (70) (1951) 567-588 (Amer. Math.Soc. Translations (2), vol. 2).

[17] Mitchell, B.: Theory of categories; Academic Press 1965.

[18] Murasugi, K.: The center of a group with a single defining relation, Math. Annalen 155 (1964) 246-251.

[19] Newman, B.B.: Some results on one-relator groups, Bull. Amer. Math. Soc. 74 (1968) 568-571.

[20] Osofsky, B.L.: Homological dimension and the continuum hypothesis, Trans. Amer. Math. Soc. 132 (1968) 217-230.

[21] Papakyriakopoulos, C.D.: On Dehn's lemma and the asphericity of knots, Annals of Math. 66 (1957) 1-26.

[22] Schur, I.: Über die Darstellung der symmetrischen und der alternierenden Gruppe durch gebrochene lineare Substitutionen, Crelle 139 (1911) 155-250.

[23] Serre, J.-P.: Sur la dimension cohomologique des groupes profinis, Topology 3 (1965) 413-420.

[24] Serre, J.-P.: Cohomologie des groupes discrets, C.R. Acad. Sc. Paris 268 (1969) 268-271.

[25] Stallings, J.: Centerless groups - an algebraic formulation of Gottlieb's theorem, Topology 4 (1965) 129-134.

[26] Stallings, J.: Homology and central series of groups, J. of Algebra 2 (1965) 170-181.

[27] Stallings, J.: Groups of dimension one are locally free, Bull. Amer. Math. Soc. 74 (1968) 361-364.

[28] Stammbach, U.: Anwendungen der Homologietheorie der Gruppen auf Zentralreihen und auf Invarianten von Präsentierungen, Math. Zeit. 94 (1966) 157-177.

[29] Stammbach, U.: Ein neuer Beweis eines Satzes von Magnus, Proc. Cambridge Phil. Soc. 63 (1967) 929-938.

[30] Swan, R.G.: The non-triviality of the restriction map in the cohomology of groups, Proc. Amer. Math. Soc. 11 (1960) 885-887.

[31] Swan, R.G.: Minimal resolutions for finite groups, Topology 4 (1965) 193-208.

[32] Swan, R.G.: Groups of cohomological dimension one, J. of Algebra 12 (1969) 585-610.

[33] Varadarajan, K.: Dimension, category and $K(\pi,n)$-spaces, J. Math. and Mech. 10 (1961) 755-771.

[34] Wolf, J.A.: Spaces of constant curvature; McGraw-Hill, 1967.

CHAPTER 9

EXTENSION CATEGORIES : GENERAL THEORY

9.1 The categories $\left(\frac{G}{-}\right)$ and $\mathcal{Q}_G$.

Classical extension theory is concerned with the extensions by a fixed group G of a fixed G-module A. We propose now to begin the study of a somewhat more general situation in which we still fix G but consider <u>all</u> extensions by G with abelian kernels. This may be expected to yield information about G itself as opposed merely to the pair (G,A).

Let G be a given group. We denote by $\left(\frac{G}{-}\right)$ the category whose objects are all extensions

$$1 \longrightarrow A \longrightarrow E \longrightarrow G \longrightarrow 1 \qquad (1)$$

where A is a G-module, and in which morphisms are pairs of group homomorphisms (α,σ) so that

$$\begin{array}{ccccccccc} 1 & \longrightarrow & A & \longrightarrow & E & \longrightarrow & G & \longrightarrow & 1 \\ & & \downarrow{\scriptstyle\alpha} & & \downarrow{\scriptstyle\sigma} & & \downarrow{\scriptstyle=} & & \\ 1 & \longrightarrow & A_1 & \longrightarrow & E_1 & \longrightarrow & G & \longrightarrow & 1 \end{array}$$

commutes. We usually abbreviate (1) as $(A|E)$ and often write (α,σ) as (σ).

Remark. The notation $\binom{G}{-}$ conflicts with that used briefly at the beginning of chapter 5 (p.65): we there wrote $\binom{G}{-}$ for the category of all extensions by G (with unrestricted kernel). There should, however, be no confusion because from now on only extensions with abelian kernel will be considered. Perhaps the proper degree of generality would be to study the category of all extensions by G with kernels in some given class of groups closed under subgroups, quotient groups and finite direct products. Some of the results below would apply in this more general situation. But the passage to the second category $\mathfrak{D}_G$ (see below) would involve overcoming certain difficulties.

Elementary consequences:

(i) If (α,σ) is a morphism, α is necessarily a G-module homomorphism.

(ii) Given (α,σ), α is surjective (injective) if, and only if, σ is surjective (injective).

(iii) $(A|E)$, $(A|E_1)$ are equivalent extensions (p.66) if there exists $(1,\sigma) : (A|E) \longrightarrow (A|E_1)$.

Recall that each object in $\binom{G}{-}$ gives rise to a 2-cohomology class as follows. Take any free presentation $1 \to R \to F \xrightarrow{\pi} G \to 1$, and lift π to $\theta : F \longrightarrow E$:

$$\begin{array}{rcccccccc} & 1 & \longrightarrow & R & \longrightarrow & F & \longrightarrow & G & \longrightarrow 1 \\ & & & & & \downarrow{\scriptstyle\theta} & & \downarrow{\scriptstyle =} & \\ (A|E): & 1 & \longrightarrow & A & \longrightarrow & E & \longrightarrow & G & \longrightarrow 1 \end{array} \qquad (2)$$

Then θ_R (= restriction of θ to R) is in $\mathrm{Hom}_F(R,A)$ and this yields the element of $H^2(G,A)$, call it co(A|E), corresponding to (A|E), by the basic Theorem 1 of chapter 5 (p.71).

Suppose $(\alpha,\sigma) : (A|E) \rightarrow (A_1|E_1)$. Then $\theta_R\alpha$ is in $\mathrm{Hom}_F(R, A_1)$ and this determines $co(A_1|E_1)$. Hence

$$\alpha^* = H^2(G,\alpha): \quad co(A|E) \rightarrow co(A_1|E_1). \qquad (3)$$

We are now prompted to introduce a second category, call it $\mathfrak{Q}_G$. The objects are all pairs (A,x), where $A \in \mathrm{Mod}_G$ (= category of G-modules) and $x \in H^2(G,A)$; and morphism

$$\alpha : (A,x) \rightarrow (A_1,x_1)$$

means a module homomorphism $\alpha : A \rightarrow A_1$ so that $x\alpha^* = x_1$. ($\mathfrak{Q}_G$ is the category of $H^2(G,\ )$-pointed objects of Mod_G : see MacLane [7], p.53).

For each (A|E), put $\Gamma(A|E) = (A, co(A|E))$. If (α,σ) is a morphism in $\left(\frac{G}{\ }\right)$, then α is a morphism in $\mathfrak{Q}_G$, by (3) above. So if we put $\Gamma(\alpha,\sigma) = \alpha$, then Γ is a functor: $\left(\frac{G}{\ }\right) \rightarrow \mathfrak{Q}_G$.

Definition. Given categories $\mathfrak{A}, \mathfrak{L}$ and a functor $F: \mathfrak{A} \rightarrow \mathfrak{L}$. We call F surjective (= full, representative) if

(i) every $\mathfrak{L}$-object has the form AF, for some A in $\mathfrak{A}$; and

(ii) F maps $\mathfrak{A}(A_1,A_2)$ onto $\mathfrak{L}(A_1F, A_2F)$.

THEOREM 1. (Surjectivity Theorem). $\Gamma: \left(\frac{G}{\ }\right) \rightarrow \mathfrak{Q}_G$ is a surjective functor.

Proof. It only remains to prove the following: Given

$$\alpha : \Gamma(A|E) \longrightarrow \Gamma(A_1|E_1),$$

then there exists σ: $E \longrightarrow E_1$ such that

$$(\alpha,\sigma) : (A|E) \longrightarrow (A_1|E_1).$$

Choose θ as in (2) above so that $\theta_R \in co(A|E)$ and $\theta_R\alpha \in co(A_1|E_1)$. We assert there exists $\varphi: F \longrightarrow E_1$ lifting $F \longrightarrow G$ such that $\varphi_R = \theta_R\alpha$. For take any ψ: $F \longrightarrow E_1$ lifting $F \longrightarrow G$ and get ψ_R, $\theta_R\alpha$ cohomologous: so $\psi_R = \theta_R\alpha + d$, for some d in $Der(F,A_1)$. Hence φ: $w \longrightarrow (w\psi)(wd)^{-1}$ will do.

Every element in E can be written (but not necessarily uniquely) as $e = (w\theta)a$ and so we define σ by $e \longrightarrow (w\varphi)(a\alpha)$. This is unambiguous, σ is a homomorphism and $\sigma = \alpha$ on A.

<u>Remark</u>. If (σ_1), (σ_2) are two morphisms such that $\sigma_1 = \sigma_2$ on A, then $\Gamma(\sigma_1) = \Gamma(\sigma_2)$. We therefore lose some group theoretical information in passing from $\left(\frac{G}{-}\right)$ to $\mathfrak{Q}_G$. Nevertheless in much of our theory it will be sufficient (and simpler, in fact) to work in $\mathfrak{Q}_G$.

Suppose H is a second group and $\theta: H \longrightarrow G$ is a given homomorphism. Given $(A|E)$ in $\left(\frac{G}{-}\right)$, we construct the pull-back for

$$\begin{array}{ccccccccc} & & & & & & H & & \\ & & & & & & \downarrow\theta & & \\ (A|E): & 1 & \longrightarrow & A & \longrightarrow & E & \xrightarrow{\pi} G & \longrightarrow & 1 \end{array}$$

and call it $(A|E)\theta'$. Thus $(A|E)\theta'$ is the extension

$$1 \longrightarrow A \longrightarrow \tilde{E} \longrightarrow H \longrightarrow 1,$$

where $\tilde{E}$ is the subgroup of $E \times H$ consisting of all (e,h) so that $e\pi = h\theta$. If (α,σ): $(A|E) \longrightarrow (A_1|E_1)$, then $(e,h) \longmapsto (e\sigma,h)$ provides a morphism $(\alpha,\sigma)\theta' : (A|E)\theta' \longrightarrow (A_1|E_1)\theta'$. Thus θ'

is a functor: $\left(\underline{G}\right) \to \left(\underline{H}\right)$.

The homomorphism θ also gives the lifting (cf. p.89),

$$\theta^*_M : H^*(G,M) \to H^*(H, M\rho),$$

where $\rho : \mathrm{Mod}_G \to \mathrm{Mod}_H$ is determined by θ. For each (A,x) in $\mathfrak{Q}_G$, let $(A\rho, x\theta^2_A) = (A,x)\theta''$. If $\alpha : (A,x) \to (B,y)$, then also $\alpha : (A,x)\theta'' \to (B,y)\theta''$ (because lifting is a homomorphism of functors). So θ'' is a functor: $\mathfrak{Q}_G \to \mathfrak{Q}_H$.

The functors θ', θ'' commute with Γ: i.e.,

$$\begin{array}{ccc} \left(\underline{G}\right) & \xrightarrow{\theta'} & \left(\underline{H}\right) \\ \Gamma\big\downarrow & & \big\downarrow\Gamma \\ \mathfrak{Q}_G & \xrightarrow{\theta''} & \mathfrak{Q}_H \end{array}$$

is commutative.

9.2 Two theorems of Schur.

As an immediate application of Theorem 1 we prove two important results in group theory, both due to Schur.

PROPOSITION 1. If $\mathrm{co}(A|E)$ has finite order n, then $(A|E)$ admits a morphism (n,σ) into the split extension $(A|S)$.

Proof. $\mathrm{co}(A|S) = 0$ and $n\,\mathrm{co}(A|E) = 0$. So n is a morphism: $\Gamma(A|E) \to \Gamma(A|S)$. Therefore by Theorem 1 there exists σ, as required.

Since σ is injective or surjective according as n is, we have

COROLLARY 1. *If co(A|E) has order n and $A \xrightarrow{n} A$ is an automorphism, then (A|E) is split.*

In particular, if A and G are finite of coprime orders, then every extension of A by G splits. (Schur).

COROLLARY 2. *If G is finite and A is torsion-free, then every (A|E) can be embedded in the split extension of A by G.*
(This result is needed in the theory of Bieberbach groups: cf. [11], chapter 3.)

COROLLARY 3. *If co(A|E) has order n and A is G-trivial, then $(A \cap E')^n = 1$.*

Proof. $(n,\sigma) : (A|E) \to (A|S)$ clearly implies $(A\cap E')\sigma \leq A\cap S'$ and $A\cap S' = 1$ because $A = A^G$. But $(A\cap E')\sigma = (A\cap E')^n$.

Corollary 3 yields immediately

THEOREM 2. (Schur) *If $E/\zeta_1(E)$ has order n, then E' is finite and $(E')^{n^2} = 1$.*

Proof. The hypothesis implies E' is finitely generated (because there are only a finite number of distinct commutators) and if $A = \zeta_1(E)$, $E'/E' \cap A$ is finite of exponent dividing n. So $E' \cap A$ is finitely generated and by Corollary 3, it has exponent dividing n.

We mention three other proofs of Theorem 2:

(1) By transfer: e.g. P. Hall's Canadian notes, §8, or Huppert's book, p.417.

(2) By a completely elementary (but not easily remembered) argument due to D. Ornstein: Kaplansky [6], p.59.

(3) Take $1 \to S \to F \xrightarrow{\pi} E \to 1$ with F free and let $R = \zeta_1(E)\pi^{-1}$. So $S \geq [R,F]$ and it is enough to prove the result for $F/[R,F]$. Now $F'/R \cap F' \cong G'$ (where $G = F/R$) and is therefore finite of exponent dividing n. As before, $R \cap F'$ is finitely generated modulo $[R,F]$. Finally, $n(R \cap F') \leq [R,F]$ because $n\, H_2(G,\mathbb{Z}) = 0$ (this being the homology analogue of Corollary 1 on p.91).

For further results concerning Theorem 2 see Baer [2], P. Hall [5], R. Turner-Smith [10].

9.3 Monomorphisms and epimorphisms.

PROPOSITION 2. The following are equivalent:

(i) (α,σ) is a monomorphism (epimorphism) in $(\underline{G})$;

(ii) α is a monomorphism (epimorphism) in $\mathfrak{Q}_G$;

(iii) α is an injective (surjective) homomorphism in Mod_G.

Both (iii) ⇒ (i) and (iii) ⇒ (ii) are trivial.

Proofs of (ii) ⇒ (iii) (D.E. Cohen):

Mono: Given a monomorphism $\alpha : (A,x) \longrightarrow (A_1,x_1)$, let $B = \mathrm{Ker}\alpha$.

Define morphisms α_1, α_2:

$$(B \oplus A,\ 0 + x) \longrightarrow (A,x)$$

by $\alpha_1: (b,a) \longmapsto a$;

$\alpha_2: (b,a) \longmapsto a+b$.

So $\alpha_1\alpha = \alpha_2\alpha$, therefore $\alpha_1 = \alpha_2$ and so $B = 0$.

Epi: Given an epimorphism $\alpha : (A,x) \longrightarrow (A_1,x_1)$, let $C = \mathrm{Im}\alpha$.

Define α_1, α_2 :

$$(A_1,x_1) \longrightarrow (\ (A_1/C) \oplus A_1,\ 0+x_1) .$$

by $\alpha_1 : a_1 \longmapsto (a_1+C,\ a_1)$;

$\alpha_2 : a_1 \longmapsto (0,\ a_1)$.

(Note that α_1 is indeed a morphism because if $\theta: R/R' \longrightarrow A$ gives x, then $\theta\alpha\alpha_1$ is zero in the first component.) Now $\alpha\alpha_1 = \alpha\alpha_2$, therefore $\alpha_1 = \alpha_2$ and so $C = A_1$, as required.

Proofs of (i) ⇒ (iii):

Mono: Given a monomorphism $(\alpha,\sigma) : (A|E) \longrightarrow (A_1|E_1)$, let $B = \mathrm{Ker}\alpha = \mathrm{Ker}\sigma$ and form $E_2 = E[B$ (the split extension).

Define homomorphisms $E_2 \longrightarrow E$ by

$$\sigma_1 : (e,b) \longmapsto e;$$

$$\sigma_2 : (e,b) \longmapsto eb.$$

Clearly $\sigma_1\sigma = \sigma_2\sigma$ and so $(\sigma_1)(\sigma) = (\sigma_2)(\sigma)$. As (σ) is monomorphic $(\sigma_1) = (\sigma_2)$ and so $B = 1$, as required.

Epi: Given an epimorphism $(\alpha,\sigma) : (A|E) \to (A_1|E_1)$, let $C = \operatorname{Im}\alpha$ and suppose

$$((A_1/C) \oplus A_1,\ 0 + co(A_1|E_1)) = \Gamma(A_2|E_2).$$

Take α_1 as in the proof of (ii) ⇒ (iii) above, and use Theorem 1 to get σ_1 so that $(\alpha_1,\sigma_1) : (A_1|E_1) \to (A_2|E_2)$. Now we have

$$\begin{array}{ccc} (A|E) & \xrightarrow{(\alpha,\sigma)} & (A_1|E_1) \\ \Big\downarrow{\scriptstyle (\alpha\alpha_1,\sigma\sigma_1)} & & \\ (A_2|E_2) & & \end{array}$$

and this can be completed by α_2 (defined as in (ii) ⇒ (iii) above) at the module level. By the lemma below, there exists σ_2 so that (α_2,σ_2) completes the above in $\left(\underline{G}\right)$. Hence $(\sigma)(\sigma_1) = (\sigma)(\sigma_2)$, so $(\sigma_1) = (\sigma_2)$ and hence $C = A_1$ as required.

<u>LEMMA 1. (Surjectivity of Γ on triangles.) Given</u>

$$\begin{array}{ccc} (A|E) & \xrightarrow{(\alpha_1,\sigma_1)} & (A_1|E_1) \\ {\scriptstyle (\alpha_2,\sigma_2)}\Big\downarrow & & \\ (A_2|E_2) & & \end{array} \qquad (1)$$

<u>and $\alpha : A_1 \to A_2$ so that</u>

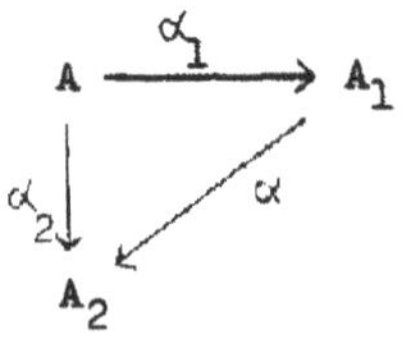

commutes. Then there exists $\sigma\colon E_1 \to E_2$ so that (α,σ) is a morphism and completes the triangle (1) in $\underline{G}$.

Proof. Since $\alpha_1{}^*\alpha^* = \alpha_2{}^*$, α does provide a morphism $(A_1,x_1) \to (A_2,x_2)$, where $x_i = \mathrm{co}(A_i|E_i)$. By the Surjectivity Theorem, there exists $\sigma\colon E_1 \to E_2$ such that $\Gamma(\alpha,\sigma) = \alpha$.

If $\sigma_1\sigma = \sigma_2$, done. If not, $e\sigma_1\sigma = e\sigma_2 . ed$, for some derivation d of E in A_2 vanishing on A (because $\alpha_1\alpha = \alpha_2$). So d is really a derivation of G in A_2. Map $E_1 \to E_2$ by

$$\sigma' : e_1 \mapsto e_1\sigma(ed)^{-1},$$

where e is any element in E such that e, e_1 have the same image in G. Then σ' is a homomorphism extending α and $\sigma_1\sigma' = \sigma_2$, as required.

9.4 Injective objects.

An object B in a category $\mathfrak{A}$ is called injective if every diagram

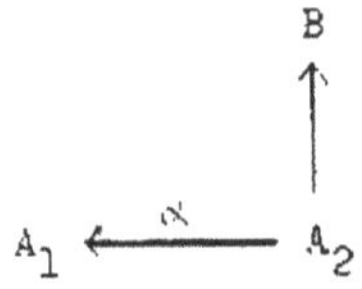

with α a monomorphism, can be completed by a morphism $\tau : A_1 \to B$.

Similarly, B is *projective* if we reverse the arrows in the above definition and assume α is epimorphic.

THEOREM 3. *The following are equivalent conditions on $(A|E)$:*

(i) $(A|E)$ is injective in $\left(\underline{G}\right)$;

(ii) $\Gamma(A|E)$ is injective in $\mathcal{Q}_G$;

(iii) A is an injective G-module (and so $co(A|E) = 0$, i.e., $(A|E)$ is split).

Proof. *(i) ⇒ (ii)*: We are required to complete

$$\begin{array}{ccc} & & \Gamma(A|E) \\ & & \uparrow{\scriptstyle \alpha_1} \\ \Gamma(A_2|E_2) & \xleftarrow[\text{mono } \alpha_2]{} & \Gamma(A_1|E_1) \end{array}$$

Let $\alpha_2 = \Gamma(\alpha_2, \sigma_2), \alpha_1 = \Gamma(\alpha_1, \sigma_1)$ (Theorem 1). By hypothesis, there exists (α, σ) completing the diagram in $\left(\underline{G}\right)$. Hence $\Gamma(\alpha, \sigma)$ completes the given diagram.

(ii) ⇒ (iii): Embed A in an injective G-module I: $i : A \to I$. Then i gives a monomorphism: $(A,x) \to (I,y)$, where $x = co(A|E)$ and $y = xi^*$.

If (A,x) is injective, there exists $j: (I,y) \to (A,x)$ such that ij is the identity on (A,x). Therefore A is a direct summand of I and so G-injective.

(Of course $y=0$ because $H^2(G,I) = 0$. We have not used

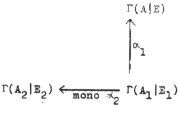

this because we wish the argument to apply in more general situations: cf. §9.9.)

(iii) ⇒ (i): Given

$$\begin{array}{ccc} & & (A|E) \\ & & \uparrow{\scriptstyle (\alpha_1,\sigma_1)} \\ (A_2|E_2) & \xleftarrow[\text{mono } (\alpha_2,\sigma_2)]{} & (A_1|E_1) \end{array}$$

with A injective. There exists $\alpha: A_2 \to A$ completing the module triangle. Hence there exists (α,σ) completing the given triangle, by Lemma 1 (§9.3, p.193).

COROLLARY. $(\underline{G})$, $\mathcal{D}_G$ both "have enough injectives".

(This means that each object admits a monomorphism into an injective.)

Proof. Take (A,x) and a module monomorphism $i: A \to I$ with I injective. Then $i: (A,x) \to (I,0)$ is a monomorphism and if $\Gamma(A|E) = (A,x)$, $S = G[I$ (split extension), $(A|E) \to (I|S)$ is also a monomorphism.

9.5 Projective objects.

Given (A|E) and a subset S of E generating E, we shall say S generates (A|E). A "map" $\varphi: S \to (A_1|E_1)$ means a mapping

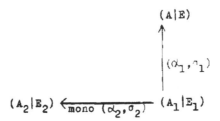

φ: $S \longrightarrow E_1$ so that
$$\begin{array}{ccc} S & \xrightarrow{\varphi} & E_1 \\ \downarrow & & \downarrow \\ E & \longrightarrow & G \end{array}$$
commutes.

We say $(A|E)$ is __freely generated by__ S (is __free on__ S) if S generates $(A|E)$ and every map φ: $S \longrightarrow (A_1|E_1)$ extends to a morphism $(A|E) \longrightarrow (A_1|E_1)$. $(A|E)$ is free if it is free on some S.

__PROPOSITION 3. The free objects in $\binom{G}{-}$ are all of the form $(\bar{R}|\bar{F})$, where $1 \longrightarrow \bar{R} \longrightarrow \bar{F} \longrightarrow G \longrightarrow 1$ is exact, F is free and bars denote things modulo R'.__

__Proof.__ Clearly all objects $(\bar{R}|\bar{F})$ are free (on $\bar{S}$ where S freely generates F).

Suppose $(A|E)$ is free on S. Take a set X in one-one correspondence with S : $\sigma_o : X \xrightarrow{\sim} S$. If F is free on X, σ_o extends to a surjective homomorphism σ: $F \longrightarrow E$. Hence we get $\bar{\sigma} : \bar{F} \longrightarrow E$. Let $R = A\sigma^{-1}$ and suppose $(\bar{R}|\bar{F})$: $1 \longrightarrow \bar{R} \longrightarrow \bar{F} \longrightarrow G \longrightarrow 1$ is the extension obtained from $\bar{F} \longrightarrow E \longrightarrow G$. Then we have an epimorphism $(\bar{\sigma})$: $(\bar{R}|\bar{F}) \longrightarrow (A|E)$. By the freeness of $(A|E)$ on S and since σ_o^{-1} gives a "map" : $S \longrightarrow (\bar{R}|\bar{F})$, there exists (α_1,σ_1) extending this: $(A|E) \longrightarrow (\bar{R}|\bar{F})$. Then (α_1,σ_1) is an epimorphism as X generates F. Also $\sigma_1\sigma = 1_E$, so that σ_1 is injective. Hence σ_1 - and therefore also (α_1,σ_1) - is an isomorphism.

Obviously every free object is projective. Also, by

Theorem 1, $(A|E)$ projective implies $\Gamma(A|E)$ projective. Hence all objects $(\bar{R},\chi)$, where $\chi = co(\bar{R}|\bar{F})$, the characteristic class of $(\bar{R}|\bar{F})$, are projective in $\mathfrak{Q}_G$. Consequently we have the

COROLLARY. $\left(\frac{G}{}\right)$, $\mathfrak{Q}_G$ both "have enough projectives".

(This means that every object is an epimorphic image of a projective.)

THEOREM 4. The following are equivalent conditions on $(A|E)$:

(i) $(A|E)$ is projective in $\left(\frac{G}{}\right)$;

(ii) $\Gamma(A|E) = (A,x)$ is projective in $\mathfrak{Q}_G$;

(iii) there exists a free object $(\bar{R}|\bar{F})$ and a G-module isomorphism $\varphi: \bar{R} \xrightarrow{\sim} A \oplus P$ so that $\varphi^*: \chi = co(\bar{R}|\bar{F}) \to (x,0)$;

(iv) there exists a free object $(\bar{R}|\bar{F})$ and $P \lhd \bar{F}$, $P \leq \bar{R}$ so that $\bar{F}$ splits over P and $(A|E) \simeq (\bar{R}/P|\bar{F}/P)$.

Remark. It is easy to see that $\mathfrak{Q}_G$ has products. Hence (iii) can be expressed thus: $(\bar{R},\chi) \simeq (A,x) \sqcap (P,0)$.

Proof. (i) ⇒ (ii) is exactly like (i) ⇒ (ii) of Theorem 3.

(ii) ⇒ (iii): There exists a free object $(\bar{R}|\bar{F})$ and an epimorphism $(\sigma) : (\bar{R}|\bar{F}) \to (A|E)$. Then σ_R is an epimorphism $(\bar{R},\chi) \to (A,x)$. The projectivity of (A,x) yields $\psi: (A,x) \to (\bar{R},\chi)$ so that $\psi\sigma_R$ is the identity on (A,x). So

$$\bar{R} = A\psi \oplus P,$$

where $P = \text{Ker}\,\sigma$ and $x\psi^* = \chi$. It remains to see that if ν is the projection $\bar{R} \to P$, then $\chi\nu^* = 0$. This is a consequence of

LEMMA 2. *Given an epimorphism $\alpha : (A,y) \to (A_1,y_1)$, where (A_1,y_1) is projective, then*

$$A \overset{\cong}{\to} A_1 \oplus B, \text{ where } B = \operatorname{Ker}\alpha$$

and

$$y \mapsto (y_1, 0).$$

Proof. By the projectivity of (A_1,y_1), there exists β so that

$$0 \to A_1 \xrightarrow{\beta} A \xrightarrow{\gamma} B \to 0$$

is exact and $y_1\beta^* = y$.

The exactness of the cohomology sequence shows that

$$H^2(A_1) \xrightarrow{\beta^*} H^2(A) \xrightarrow{\gamma^*} H^2(B)$$

is exact and hence $y\gamma^* = y_1\beta^*\gamma^* = 0$.

For (iii) ⇒ (iv) we use

LEMMA 3. *Given $(A|E)$, let N be a direct G-summand of A, ν the projection $A \to N$ and θ the projection $E/N \to G$. If $x = co(A|E)$, then $x\nu^*\theta_N^2$ is the cohomology class of*

$$1 \to N \to E \to E/N \to 1.$$

Proof. Let M be the kernel of ν. Then $(N, x\nu^*) = \Gamma(N|E/M)$

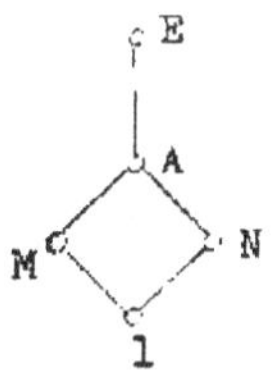

and $x\nu^*\theta_N^2$ is the cohomology class of the pull-back for

$$\begin{array}{ccccccccc} & & & & & & E/N & & \\ & & & & & & \downarrow & & \\ 1 & \to & N & \to & E/M & \to & G & \to & 1 \end{array}$$

(cf. end of §9.1, p.189). If E_1 is the pull-back extension, then $e \mapsto (eM, eN)$ is an isomorphism of E onto E_1.

<u>Proof of (iii) ⇒ (iv):</u> The epimorphism $(\bar{R}|\bar{F}) \longrightarrow (A|E)$ gives $(\bar{R}/P|\bar{F}/P) \xrightarrow{\cong} (A|E)$.

If ν is the projection $\bar{R} \longrightarrow P$, θ is the projection $F/P \longrightarrow G$, then $\chi\nu^*\theta_P^2 = 0$, because $\chi\nu^* = 0$, and by Lemma 3, this is the class of $(P|\bar{F})$ in the category $\left(\frac{\bar{F}/P}{}\right)$.

<u>(iv) ⇒ (i):</u> If $(A|E)$ is as in (iv), $\bar{F} = E[P$. Consider

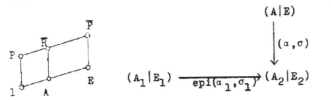

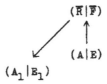

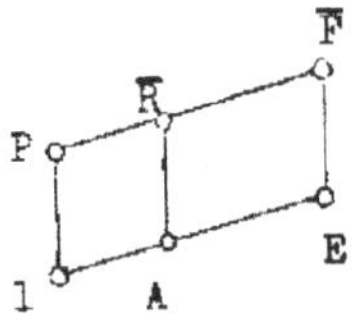

$$\begin{array}{ccc} & & (A|E) \\ & & \downarrow{\scriptstyle (\alpha,\sigma)} \\ (A_1|E_1) & \xrightarrow[\text{epi}(\alpha_1,\sigma_1)]{} & (A_2|E_2) \end{array}$$

By hypothesis there exist morphisms $(\bar{R}|\bar{F}) \leftrightarrows (A|E)$ so that the appropriate product is the identity on $(A|E)$. By the freeness of $(\bar{R}|\bar{F})$, there exists $(\bar{R}|\bar{F}) \longrightarrow (A_1|E_1)$ completing the appropriate triangle. Then

$$\begin{array}{ccc} & & (\bar{R}|\bar{F}) \\ & \swarrow & \uparrow \\ & & (A|E) \\ (A_1|E_1) & & \end{array}$$

will complete the original triangle.

The module P in part (iii) of Theorem 4 is necessarily G-projective. This is an immediate consequence of the following result.

PROPOSITION 4. *If (A,x) is projective and*

$$A = A_1 \oplus B$$

$$x = (x_1, 0),$$

then B is a projective G-module.

Proof. Consider a diagram of G-modules:

$$\begin{array}{ccc} & & B \\ & & \downarrow{\scriptstyle\beta} \\ C_1 & \xrightarrow[\text{epi}\,\gamma]{} & C_2 \end{array}$$

This yields in $\mathcal{Q}_G$ the diagram:

$$\begin{array}{ccc} & & (B,0) \\ & & \downarrow{\scriptstyle\beta} \\ (C_1,0) & \xrightarrow[\text{epi}\,\gamma]{} & (C_2,0). \end{array}$$

If $\nu\colon A \longrightarrow B$ is the natural projection, $x\nu^* = 0$ and so $\nu\beta : (A,x) \longrightarrow (C_2,0)$ yields (by the projectivity of (A,x)) a morphism $\alpha : (A,x) \longrightarrow (C_1,0)$. So $\alpha\gamma = \nu\beta$ on the modules and restricting α to B we get the required mapping completing the original triangle.

9.6 Minimal projectives.

Given (A,x) projective and P a projective G-module, it is easy to verify that $(A \oplus P, (x,0))$ is also projective. (We leave this as an exercise: Enlarge P to a free module V and if (A,x) comes from $(\bar{R},\chi)$, prove that $(\bar{R} \oplus V, (\chi,0))$ is also free.)

We thus see that one can always produce new projectives in a very simple manner from given ones. To compare arbitrary projectives does not therefore seem very profitable. The situation is quite different if we restrict attention to those projectives that do not admit a decomposition as in Proposition 4.

<u>Definition</u>. The projective extension $(A|E)$ is called <u>minimal</u> if $A \neq 0$ and whenever $(\sigma): (A|E) \longrightarrow (A_1|E_1)$ is an epimorphism with $(A_1|E_1)$ projective, then (σ) is an isomorphism.

We make a similar definition in the category $\mathfrak{Q}_G$.

<u>Exercise</u>. $(A|E)$ is minimal if, and only if, $\Gamma(A|E)$ is minimal.

PROPOSITION 5. <u>The projective (A,x) is minimal if, and only if, there does not exist a splitting</u>

$$\underline{A = A_1 \oplus B}$$
$$\underline{x = (x_1,0)}$$

<u>with $B \neq 0$.</u>

<u>Proof</u>. Assume such a splitting. By Theorem 4, (ii) ⇔ (iii), (A_1,x_1) is projective and the epimorphism $(A,x) \longrightarrow (A_1,x_1)$ is not an isomorphism.

Conversely, let $\alpha : (A,x) \longrightarrow (A_1,x_1)$ be an epimorphism with (A_1,x_1) projective. Then (Lemma 2)

$$A \xrightarrow{\simeq} A_1 \oplus B$$
$$x \longmapsto (x_1,0)$$

where B = Kerα. If we assume there is no such splitting, B = 0 and therefore α is an isomorphism.

By Proposition 4, the projective (A,x) is minimal if A has no G-projective direct summand. If G is finite, by p.22,

$$H^2(G, \text{projective}) = 0$$

and hence we have the

<u>COROLLARY. If G is finite, the projective (A,x) is minimal if, and only if, A has no projective direct summand.</u>

A given group G possesses minimal projectives provided Mod_G contains modules $\bar{R}$ with ascending chain condition on projective direct summands.

This is true, in particular, if $\mathbb{Z}G$ has maximal condition on right ideals. For then G is finitely generated (because $\mathfrak{g}$ is finitely generated: actually G has the maximal condition on subgroups) and so there exists a finitely generated free extension $(\bar{R} | F)$. Now

$$\bar{R} \simeq \mathfrak{r}/\mathfrak{f}\mathfrak{r} \hookrightarrow \mathfrak{f}/\mathfrak{f}\mathfrak{r}$$

and $\mathfrak{f}/\mathfrak{f}\mathfrak{r}$ is finitely generated and G-free. Hence $\mathfrak{f}/\mathfrak{f}\mathfrak{r}$, and therefore also $\bar{R}$, satisfies the ascending chain condition on submodules.

<u>Problem</u>. Give a good description of the class of all groups possessing minimal projectives. Is this conceivably all groups?

9.7 Change of coefficient ring.

If K is a commutative ring, let $\mathfrak{Q}_{KG}$ be the category of pairs (A,x), where $A \in \text{Mod}_{KG}$ and morphism $\alpha : (A,x) \to (A_1,x_1)$ means a KG-homomorphism of modules $\alpha : A \to A_1$ so that $x\alpha^* = x_1$.

Let $\binom{KG}{}$ be the corresponding category of extensions $(A|E)$ where A is a KG-module and a morphism (α,σ) is a pair of group homomorphisms as before but where now α is a K-homomorphism as well (and therefore a KG-homomorphism).

We review how the above theory would look with K in place of $\mathbb{Z}$.

The Surjectivity Theorem in §9.1 remains true.

Monomorphisms and epimorphisms are characterised as before: i.e., the analogue to Proposition 2, §9.3, is true. Note that in proving Proposition 2, (i) ⇒ (iii) for monomorphisms, σ_1, σ_2 restrict to KG-module homomorphisms.

Theorem 3 (§9.4) remains true, but of course injective G-module now becomes injective KG-module. (We also recall that $H^2(G,I) = 0$ if I is KG-injective.)

For A in Mod_G, put $A_{(K)} = K \otimes_{\mathbb{Z}} A$ in Mod_{KG}. Let $i : A \to A_{(K)}$ be $a \mapsto 1 \otimes a$. If $x \in H^2(G,A)$, write $x_{(K)} = xi^*$ and

$$(A,x)_{(K)} = (A_{(K)}, x_{(K)}).$$

Then $(A,x) \to (A,x)_{(K)}$ is a functor: $\mathfrak{Q}_G \to \mathfrak{Q}_{KG}$.

<u>PROPOSITION 6. Every (A,x) in $\mathfrak{Q}_{KG}$ is an epimorphic image of some $(\bar{R},\chi)_{(K)}$.</u>

<u>Proof</u>. Let $(A,x) = \Gamma(A|E)$, Y generate A over KG, and S generate E modulo A. If S', Y' are disjoint sets in one-one correspondence with S,Y, respectively, and F is free on $S' \cup Y'$, then $s' \mapsto s$, $y' \mapsto y$ yields a homomorphism $\theta: F \to E$ and if $\tau: F \to G$ (surjective) is the product $F \xrightarrow{\theta} E \to G$, then $\operatorname{Ker}\tau = R$ maps under θ into A. Clearly $R\theta \geq Y$ and so $\bar{R}_{(K)}$ is mapped onto A by $\bar{\theta}_{(K)}$. Now $x = co(A|E)$ contains $\bar{\theta}_R$ (restriction of $\bar{\theta}$ to $\bar{R}$), and hence

$$\bar{\theta}_{(K)}: (\bar{R},\chi)_{(K)} \to (A,x)$$

is an epimorphism.

Let us call $(\bar{R},\chi)_{(K)}$ a <u>free pair (object)</u> in $\mathfrak{Q}_{KG}$. (Justify this terminology: cf. Proposition 3.) Obviously all such pairs are projective in $\mathfrak{Q}_{KG}$ and so, by Proposition 6, $\mathfrak{Q}_{KG}$ has enough projectives.

The analogue of Theorem 4 holds:

<u>THEOREM 4(K). The following are equivalent conditions on $(A|E)$ in $\left(\frac{KG}{-}\right)$:</u>

<u>(i) $(A|E)$ is projective in $\left(\frac{KG}{-}\right)$;</u>

<u>(ii) $\Gamma(A|E) = (A,x)$ is projective in $\mathfrak{Q}_{KG}$;</u>

<u>(iii) there exists $(\bar{R},\chi)$ and a KG-module isomorphism $\varphi : \bar{R}_{(K)} \xrightarrow{\sim} A \oplus P$ so that $\varphi^*: \chi_{(K)} \mapsto (x,0)$;</u>

<u>(iv) there exists $(\bar{R}_{(K)}|T)$ with $co(\bar{R}_{(K)}|T) = \chi_{(K)}$, and a KT-submodule P of $\bar{R}_{(K)}$ so that $(A|E) \cong (\bar{R}_{(K)}/P|T/P)$ and T splits over P.</u>

The analogue of Proposition 4 holds. Hence the submodule P in (iii) is KG-projective.

We may define minimal projectives and have then the analogues of Proposition 5 and its corollary.

<u>Remarks</u>

<u>(1)</u> Let $K = \mathbb{F}_p$. Then A in Mod_G implies $A_{(\mathbb{F}_p)} \cong A/pA$; and a solution for T in (iv) of Theorem 4 $(\mathbb{F}_p)$ is $\bar{F}/\bar{R}^p \cong F/R'R^p$. The free objects in $\binom{\mathbb{F}_p G}{\quad}$ are $(\bar{R}/\bar{R}^p \mid \bar{F}/\bar{R}^p)$.

<u>(2)</u> If G is finite and A is finitely generated over KG, then in Proposition 6 we can choose $\bar{R}$ also finitely generated over $\mathbb{Z}G$ and therefore $\bar{R}_{(K)}$ finitely generated over KG. Ditto in Theorem 4(K).

9.8 Projective Covers.

<u>Definition</u>. Given a category $\mathfrak{A}$ and $f: A \to B$ in $\mathfrak{A}$. Then f is called an <u>essential epimorphism</u> if f is an epimorphism and whenever $g: X \to A$ is such that gf is an epimorphism, then so is g.

The essential epimorphism $f: A \to B$ is called a <u>projective</u>

<u>cover</u> (of B) if A is projective in $\mathfrak{A}$.

<u>LEMMA 4. If $f_1 : P_1 \to A$, $f_2 : P_2 \to A$ are projective covers, then $P_1 \cong P_2$.</u>

<u>Proof</u> = exercise (or [9], p.88).

<u>Example</u>. If $\mathfrak{A}$ is the category of all finitely generated KG-modules, where K is some commutative ring, then $f: A \to B$ is an essential epimorphism if, and only if, $\operatorname{Ker} f \leq Fr_{KG}(A)$. (Use §7.1.)

In particular, if K is a field and G is finite, then any minimal epimorphism in the sense of §8.12, p.178, is an essential epimorphism. Moreover, every finitely generated KG-module has then a finitely generated projective cover.

More generally, we quote the following important fact (which we shall use in the last chapter) (cf. [9], p.91, for a proof):

<u>PROJECTIVE COVER THEOREM. Let I be an ideal of the commutative ring K so that K is I-complete and K/I satisfies the descending chain condition on ideals. If G is finite, then all finitely generated KG-modules have finitely generated projective covers.</u>

To say that K is I-complete means that $\bigcap I^n = 0$ and K is complete in the topology determined by (I^n) (i.e., all Cauchy sequences have a limit in K).

<u>Definition</u>. Let K be any commutative ring. We call $(A|E) \in \binom{KG}{}$ an <u>essential extension</u> if $(A|E) \longrightarrow (1|G)$ is essential in $\binom{KG}{}$ (i.e., if $AH = E$ and $A \cap H \in \mathrm{Mod}_{KG}$ always imply $H = E$).

Further, the essential $(A|E)$ is called <u>maximal</u> if whenever $(\sigma) : (A_1|E_1) \longrightarrow (A|E)$ is an epimorphism with $(A_1|E_1)$ essential, then (σ) is an isomorphism.

If $(A|E)$ is essential and projective we call it a <u>projective cover</u> (of G) (i.e., $(A|E) \longrightarrow (1|G)$ is a projective cover in $\binom{KG}{}$). Note that, by Lemma 4, any two projective covers are isomorphic.

<u>Exercises</u>.

1. What are essential extensions and projective covers in $\binom{KG}{}$ in general?

2. If G is finite, prove that every essential extension in $\binom{\mathbb{F}_p G}{}$ is finite; and further that $(A|E)$ is essential if, and only if, $A \leq \mathrm{Fr}(E)$ (= Frattini group of E).

<u>LEMMA 5. The following conditions are equivalent:</u>

<u>(i) $(A|E)$ is essential;</u>

<u>(ii) every morphism into $(A|E)$ is an epimorphism;</u>

<u>(iii) if $\alpha : \Gamma(A|E) \longrightarrow (C,0)$ is an epimorphism, then $C = 0$.</u>

<u>Proof</u>. The equivalence of (i) and (ii) is obvious.

If α is given as in (iii) and $N = \mathrm{Ker}\,\alpha$, then E/N splits over A/N: $E = AL$, $A \cap L = N$, for some L. If (i) holds, $L = E$ and so $N = A$.

Conversely, if $(B|L) \hookrightarrow (A|E)$ is an inclusion, then $(A/B|E/B)$ is split and so if (iii) holds, $A = B$.

Exercise. If G is finite and $K = \mathbb{Z}$, prove that there exist no projective covers. (Use Lemma 5(iii) and Corollary 1 to Proposition 1, p.190.)

(The situation can be very different when K is not $\mathbb{Z}$. We shall return to this existence problem in Chapter 11.)

LEMMA 6. Every projective has every essential extension as an image.

Proof. If $(A|E)$ is projective, there exists $(A|E) \hookrightarrow (\bar{R}|\bar{F})$. For any $(B|L)$, we can find $(\bar{R}|\bar{F}) \to (B|L)$ (by freeness) and so $(A|E) \to (B|L)$. If $(B|L)$ is essential, then this is necessarily an epimorphism (Lemma 5).

LEMMA 7. If $(A|E)$ is essential, $(A_1|E_1)$ is projective and

$$(\sigma) : (A|E) \to (A_1|E_1)$$

is an epimorphism, then (σ) is an isomorphism.

Proof. By the projectivity, there exists (τ); $(A_1|E_1) \to (A|E)$ so that $(\tau)(\sigma)$ is the identity. Since $(A|E)$ is essential, (τ) must be epimorphic and hence $(\tau) = (\sigma)^{-1}$.

COROLLARY. If $(B|L)$ is a projective cover, then it is minimal projective and maximal essential.

Proof. Taking $(A_1|E_1) = (B|L)$ in the lemma shows $(B|L)$ is a maximal essential extension; taking $(A|E) = (B|L)$ shows $(B|L)$ is minimal projective.

An immediate consequence of this corollary and Lemmas 4 and 6 is

PROPOSITION 7. If there exists a single projective cover, then minimal projective = projective cover = maximal essential and any two such are isomorphic.

9.9 Central extensions.

Let us see what our theory looks like when we restrict the kernels to be trivial G-modules. We shall find that we regain, in a more general setting, Schur's theory of "Darstellungsgruppen" [8].

If Tr denotes the full subcategory in Mod_G of all trivial modules, write $\left(\frac{G}{Tr}\right)$ for the corresponding category of extensions (i.e., the central extensions by G) and $\mathfrak{Q}_{TrG}$ for the related category of pairs.

Once again the Surjectivity Theorem is true; monomorphisms and epimorphisms have the same meaning as in the underlying module category; and the characterisation of injectives (Theorem 3) remains valid: $(A|E)$ is injective if, and only if, $\Gamma(A|E)$ is injective if, and only if, A is injective in Tr (i.e., A is $\mathbb{Z}$-injective = divisible). (But here of course $co(A|E)$ need not vanish.)

<u>Exercise</u>. Let $\mathcal{L}$ be a subcategory of Mod_{KG} closed under formation of (i) finite products and finite coproducts and (ii) kernels and cokernels (i.e., if $0 \to K \xrightarrow{i} A \xrightarrow{f} B \xrightarrow{p} C \to 0$ is exact in Mod_{KG} and $A \xrightarrow{f} B$ is in $\mathcal{L}$, then $K \xrightarrow{i} A$, $B \xrightarrow{p} C$ are also in $\mathcal{L}$). Prove that epimorphism and monomorphism in $\left(\frac{G}{\mathcal{L}}\right)$, $\mathfrak{D}_{\mathcal{L}G}$ have the same meaning as in Mod_{KG}.

Show further that, if $\mathcal{L}$ has enough injectives, $(A|E)$ is injective if, and only if, A is injective in $\mathcal{L}$.

If $1 \to R \to F \to G \to 1$ is a free presentation, we shall write $\tilde{}$ for things modulo $[R,F]$. Then $(\tilde{R}|\tilde{F})$ is a free extension in $\left(\frac{G}{\mathrm{Tr}}\right)$ and all free extensions are of this form (cf. Proposition 3.) The analogue of Theorem 4 reads:

<u>THEOREM 4(Tr). The following are equivalent conditions on $(A|E)$:</u>

<u>(i) $(A|E)$ is projective in $\left(\frac{G}{\mathrm{Tr}}\right)$;</u>

<u>(ii) $\Gamma(A|E) = (A,x)$ is projective in $\mathfrak{D}_{\mathrm{Tr}G}$;</u>

<u>(iii) there exists a free object $(\tilde{R}|\tilde{F})$ and a splitting $\varphi: \tilde{R} \xrightarrow{\cong} A \oplus P$ so that $\varphi^*: \tilde{\chi} = \mathrm{co}(\tilde{R}|\tilde{F}) \mapsto (x,0)$;</u>

<u>(iv) there exists $(\tilde{R}|\tilde{F})$ and $P \leq \tilde{R}$ so that $\tilde{F} = L \times P$ for some L and $(A|E) \cong (L \cap \tilde{R}|L)$.</u>

The analogue of Proposition 4 implies P in (iii) is $\mathbb{Z}$-projective (= free abelian). Minimal projectives and essential extensions are defined as before and Propositions 5 and 7 hold relative to Tr (but the corollary to Proposition 5 is false: cf. the following exercise).

<u>Exercise</u>. Let G be a finite p-group and suppose $(\tilde{R}|\tilde{F})$ comes from a minimal presentation of G (p.100). Prove that $(\tilde{R}|\tilde{F})$ is minimal projective. (Hint: Find the minimum number of generators of a split image of $(\tilde{R}|\tilde{F})$.)

So much for the previous theory. Perhaps the most striking new feature is that free extensions need not be torsion-free. In fact, <u>if G is finite,</u> then $H_2(G,\mathbb{Z}) \simeq \tilde{R} \cap \tilde{F}'$ is finite (cf. end of §9.2, p.191) and since $\tilde{R}$ splits over $\tilde{R} \cap \tilde{F}'$ (because RF'/F' is free abelian), <u>$\tilde{R} \cap \tilde{F}'$ is exactly the torsion group of $\tilde{R}$</u>.

If S is any complement to $\tilde{R} \cap \tilde{F}'$ in $\tilde{R}$, then $(\tilde{R}/S|\tilde{F}/S)$ has the property that $\tilde{R}/S$ lies in the commutator group of $\tilde{F}/S$. Such extensions are essential:

<u>LEMMA 8. If $A \leq E'$, then $(A|E)$ is essential.</u>

<u>Proof</u>. If $AH = E$, then $H' = E'$ (because A is central). Hence $A \leq H'$ by hypothesis and so $H = E$.

The converse of Lemma 8 is obviously false: take G of order p and E cyclic of order p^2.

<u>Definition</u>. If $A \leq E'$, then $(A|E)$ is called a <u>stem extension</u> and $(A|E)$ is a <u>stem pair</u> (cf. [4]).

The stem extension $(A|E)$ is called <u>maximal</u> if any epimorphism of another stem extension onto $(A|E)$ is necessarily an isomorphism.

A stem extension of the form $(\tilde{R}/S|\tilde{F}/S)$, where $(\tilde{R}|\tilde{F})$ is free and S is a complement to $\tilde{R} \cap \tilde{F}'$ in $\tilde{R}$, will be called a <u>stem cover</u> [4]; and the same term for its image under Γ. (This terminology is explained by (i) of the following proposition. Schur called these extensions <u>Darstellungsgruppen</u> for G; his term is explained by the discussion following Theorem 6 below.)

<u>PROPOSITION 8. (i) Every stem extension is the image of some stem cover.</u>

<u>(ii) Every stem cover is a maximal stem extension.</u>

<u>COROLLARY. The maximal stem extensions are precisely the stem covers.</u>

<u>Proof.</u> (i) Let $(A|E)$ be a stem extension. Choose any free $(\tilde{R}|\tilde{F})$ and an epimorphism $(\sigma) : (\tilde{R}|\tilde{F}) \longrightarrow (A|E)$. If $K = \mathrm{Ker}\sigma$, then $K + H = \tilde{R}$, where $H = \tilde{R} \cap \tilde{F}'$. Now $K/K \cap H \cong RF'/F'$ and so splits: say $K = (K \cap H) \oplus S$. Then $H \oplus S = \tilde{R}$ and $S \leq K$. Hence (σ) induces an epimorphism $(\tilde{R}/S|\tilde{F}/S) \longrightarrow (A|E)$.

(ii) In view of part (i) we have only to prove that if

$$(\sigma) : (\tilde{R}_1/S_1|\tilde{F}_1/S_1) \longrightarrow (\tilde{R}/S|\tilde{F}/S)$$

is an epimorphism of stem covers, then (σ) is an isomorphism.

If $(p) : (\tilde{R}|\tilde{F}) \longrightarrow (\tilde{R}/S|\tilde{F}/S)$ is the natural projection, then (p) can be lifted to $(\tau) : (\tilde{R}|\tilde{F}) \longrightarrow (\tilde{R}_1/S_1|\tilde{F}_1/S_1)$. By Lemma 5, (τ) is an epimorphism and if $T = \mathrm{Ker}\tau$, $T+H = \tilde{R}$. But $T \leq \mathrm{Ker}\ p = S$ (as $(\tau)(\sigma) = (p)$) and $S \oplus H = \tilde{R}$. Hence $T = S$ and (σ) is an isomorphism.

Exercise. If $H_2(G,\mathbb{Z})$ is finitely generated, then a stem extension $(A|E)$ is maximal if, and only if, $A \simeq H_2(G,\mathbb{Z})$.

<u>THEOREM 5. The following conditions are equivalent;</u>
<u>(i) There exists a single projective cover;</u>
<u>(ii) G/G' is free abelian;</u>
<u>(iii) every stem cover is projective.</u>

<u>COROLLARY. If G/G' is free abelian, then</u>

<u>minimal projective = projective cover</u>
<u>= maximal essential = stem cover</u>

<u>and any two such are isomorphic.</u>

The corollary is an immediate consequence of the theorem and Proposition 7.

Since stem covers are essential, (iii) ⇒ (i) is trivial.

<u>Proof of (i) ⇒ (ii).</u> If $(A|E)$ is projective, we can write $\tilde{F} = E \times P$ with $\tilde{R} = A \times P$ for some suitable $(\tilde{R}|\tilde{F})$. Hence $F/F' \simeq E/E' \times P$ and so E/E' is free abelian. If now $(A|E)$ is also essential, then it is easy to see that $(AE'/E'|E/E')$ is an essential extension in $\left(\frac{G/G'}{\mathrm{Tr}}\right)$ and so $AE' = E'$ or $AE' = E$. In either case G/G' is free abelian.

To prove (ii) ⇒ (iii), we first establish

<u>PROPOSITION 9. Let A be a trivial G-module so that $\mathrm{Ext}^1_{\mathbb{Z}}(G/G', A) = 0$. Then every diagram</u>

$$\begin{array}{ccc} (\tilde{R},\tilde{\chi}) & \xrightarrow{\alpha} & (A,x) \\ \downarrow{\scriptstyle p} & & \\ (H,y) & & \end{array}$$

<u>where p is a projection on a stem cover (H,y), can be completed by α_H, the restriction of α to H.</u>

<u>Proof</u>. Clearly $\alpha - p\alpha_H$ is zero on H and equals α on S = Ker p. Our hypothesis says that there exists γ: F/F' $\longrightarrow$ A completing the diagram

$$\begin{array}{ccc} S & \xrightarrow{q} & F/F' \\ {\scriptstyle \alpha_S}\downarrow & & \\ A & & \end{array}$$

where q: $\tilde{r} \longrightarrow rF'$. (We see this either by applying directly the definition of Ext^1 according to which $Ext^1(G/G',A)$ is the cokernel of Hom(F/F', A) $\longrightarrow$ Hom(S,A); or by viewing Ext^1 as Ext, the $\mathbb{Z}$-module extensions (cf. §10.1) and using the fact that the push-out of the above diagram splits by hypothesis.) Thus $\alpha - p\alpha_H$ is cohomologous to zero (cf. Proposition 6, p.47). Now for any $\beta \in Hom(\tilde{R},A)$, β lies in the cohomology class $\tilde{\chi}\beta^*$. Hence $\tilde{\chi}\alpha^* = \tilde{\chi}p^*(\alpha_H)^*$ and so

$$y(\alpha_H)^* = \tilde{\chi}p^*(\alpha_H)^* = \tilde{\chi}\alpha^* = x,$$

as required.

<u>Proof of Theorem 5, (ii) $\Rightarrow$ (iii).</u> If G/G' is free abelian, then $Ext^1_{\mathbb{Z}}(G/G', -) = 0$. We apply Proposition 9 with $(A,x) = (\tilde{R},\tilde{\chi})$ and α the identity. Then the existence of id_H: $(H,y) \longrightarrow (\tilde{R},\tilde{\chi})$ shows that $(\tilde{R},\tilde{\chi}) \cong (H,y) \sqcap (S,0)$, whence (H,y) is projective by Theorem 4(Tr).

Exercise. If G is a finite perfect group, then a minimal projective cannot be free.

Remark. It appears that we only need $\text{Ext}^1_{\mathbb{Z}}(G/G', \tilde{R}) = 0$. But then $\text{Ext}^1_{\mathbb{Z}}(G/G', S) = 0$ and so G/G' is isomorphic to a subgroup of F/F', hence is free abelian after all.

Problem. Theorem 5 ensures the existence of minimal projectives whenever G/G' is free abelian. Maximal stem covers always exist in view of Proposition 8(ii). Do minimal projectives always exist? Probably not: try the additive group of rationals (cf. the proof of Proposition 4, §8.3, p.125).

Another application of Proposition 9 is

THEOREM 6. Every stem cover admits a morphism into every injective object.

Proof. The injective pairs are precisely (A,x) where A is divisible. Hence $\text{Ext}^1_{\mathbb{Z}}(G/G', A) = 0$ and Proposition 9 applies.

An immediate corollary of Theorem 6 is an important result of Schur: if $(H|L)$ is a stem cover and ρ is a projective-geometric representation of G over an algebraically closed field F, then there exists a linear representation σ of L that induces ρ. For if we take the pull-back of

$$\begin{array}{c} \quad\quad\quad\quad\quad G \\ \quad\quad\quad\quad\quad \downarrow \\ 1 \to F^* \to GL(n,F) \to PGL(n,F) \to 1, \end{array}$$

say $(F^*|E)$, then this is injective (because F^* is divisible) and so, by Theorem 6, there exists $(\sigma_1): (H|L) \to (F^*|E)$. Then

$\sigma = \sigma_1\sigma_2$, where σ_2: $E \longrightarrow GL(n,F)$, is the required linear representation of L.

Two stem covers need not be isomorphic if G/G' is not free: for example, the two non-isomorphic non-abelian groups of order 8 are both stem covers of the Klein four-group. A weaker relationship between stem covers does however always hold.

__Definition.__ (P.Hall [4].) Two groups H, K are called __isoclinic__ ($H \sim K$) if there exist isomorphisms

$$\varphi: H/\zeta_1(H) \xrightarrow{\sim} K/\zeta_1(K),$$

$$\psi: H' \xrightarrow{\sim} K'$$

so that φ induces ψ in the following sense: if $a,b \in H$ and $\bar{a} \in a\,\zeta_1(H)\varphi$, $\bar{b} \in b\zeta_1(H)\varphi$, then $[a,b]\psi = [\bar{a},\bar{b}]$.

Note that a sufficient condition for φ to induce ψ is that there exists a third group L and surjective homomorphisms $\lambda: L \longrightarrow H$, $\mu: L \longrightarrow K$ so that

(i) $(L \xrightarrow{\lambda} H \longrightarrow H/\zeta_1(H) \xrightarrow{\varphi} K/\zeta_1(K)) = (L \xrightarrow{\mu} K \longrightarrow K/\zeta_1(K))$,

(ii) $\lambda_{L'}\psi = \mu_{L'}$.

Isoclinism is an equivalence relation on groups in which all abelian groups collapse: if H is abelian, $H \sim 1$.

__THEOREM 7. (i) All projectives lie in a single isoclinism class which also contains all stem covers.__

__(ii) Every extension is isoclinic with a stem extension.__

__Proof.__ (i) If $(\tilde{R},\tilde{\chi}) = (A,x)\pi(P,0)$, then $\tilde{F} = E \times P$ where

$\Gamma(A|E) = (A,x)$. Hence $\tilde{F} \sim E$. So we need to show that any two free objects are isoclinic.

Take $(\tilde{R}_i|\tilde{F}_i)$, $i = 1,2$. Then by Lemma 11, §8.10, p.162, $\tilde{R}_1 \times \tilde{F}_2 \simeq \tilde{R}_2 \times \tilde{F}_1$. But $\tilde{F}_2 \sim \tilde{R}_1 \times \tilde{F}_2$ and $\tilde{F}_1 \sim \tilde{R}_2 \times \tilde{F}_1$ and so $\tilde{F}_1 \sim \tilde{F}_2$.

Finally, let $(\tilde{R}/S|\tilde{F}/S)$ be any stem cover. Then $\tilde{F} \to \tilde{F}/S$ induces an isomorphism on $\tilde{F}'$ (because $\tilde{F}' \cap S = 1$) and also an isomorphism on the central quotients (because $\zeta_1(\tilde{F}/S) = \zeta_1(\tilde{F})/S$).

(ii) (Cf. [4], p.135.) Let $p\colon V \to E/E'$ be an epimorphism with V free abelian. Take the pull-back to

$$\begin{array}{ccc} & & V \\ & & \downarrow \\ E & \longrightarrow & E/E' \end{array},$$

say P. Then $P/P' \simeq V$ and $P \sim E$. Moreover, $P \to E$ yields $P \to G$, with kernel B, say. If $B = (B \cap P') \times S$, then $P \sim P/S$ and $(B/S|P/S)$ is a stem extension.

<u>Problem</u>. Suppose G is finite but not perfect. How do the minimal projectives compare?

<u>Exercise</u>. Discuss $\left(\frac{\mathbb{F}_p G}{\mathrm{Tr}}\right)$, $\mathfrak{Q}_{\mathrm{Tr}\mathbb{F}_p G}$.

Sources and references.

The categorical approach to extension theory adopted in this chapter was also used in a work of Charlap [3].

Theorem 1 (the Surjectivity Theorem) must be well known. But the only places where I know it to be written down explicitly are on p.179 of [1] (Theorem 2) and in chapter 8 of Lang's book.

The application of Proposition 1 to Theorem 2 was shown to me by J.A. Green.

For further group theoretic applications of Schur's theory of central extensions (§9.9) we refer to Huppert's book, chapter V, §§23, 24, 25.

[1] Artin, E. and Tate, J.: Class field theory; Benjamin, New York, 1967.

[2] Baer, R.: Endlichkeitskriterien für Kommutatorgruppen, Math.Annalen 124 (1952) 161-177.

[3] Charlap, L.: Compact flat Riemannian manifolds, I, Ann. of Math. 81 (1965), 15-30.

[4] Hall, P.: The classification of prime-power groups, Crelle 182 (1940) 130-141.

[5] Hall, P.: Finite-by-nilpotent groups, Proc.Cambridge Phil.Soc. 52 (1956) 611-616.

[6] Kaplansky, I.: An introduction to differential algebra; Hermann, Paris, 1957.

[7] MacLane, S.: Categorical algebra, Bull.Amer.Math.Soc. 71 (1965) 40-106.

[8] Schur, I.: Über die Darstellung der endlichen Gruppen durch gebrochene lineare Substitutionen, Crelle 127 (1904), 20-50.

[9] Swan, R.G.: Algebraic K-theory, Springer Lecture Notes Vol. 76 (1968).

[10] Turner-Smith, R.F.: Marginal subgroup properties for outer commutator words, Proc.London Math.Soc. 14 (1964) 321-341.

[11] Wolf, J.A.: Spaces of constant curvature, McGraw-Hill 1967.

There is an extremely interesting discussion of the lower central series in a recent paper by L.Evens: Terminal p-groups (Illinois J. Math. 12(1968) 682-699). This can be efficiently expressed in the language developed in this chapter.

CHAPTER 10

MORE MODULE THEORY

Our purpose in this chapter is to collect together some results (mostly cohomological) on modules over finite groups, that are needed in the next (and last) chapter.

We shall assume throughout that all groups are finite.

Notation. If G is a (finite) group and K is a commutative ring, let Lat_{KG} denote the full subcategory of Mod_{KG} consisting of all K-projective and finitely generated KG-modules. The objects of Lat_{KG} are called KG-lattices. Observe that all short exact sequences in Lat_{KG} split over K.

10.1 Module extensions.

Let A, C be KG-modules and assume either that C is K-projective or A is K-injective. This ensures that every module extension of A by C splits over K.

Suppose we have such an extension

$$0 \longrightarrow A \longrightarrow B \xrightarrow{\pi} C \longrightarrow 0 \qquad (1)$$

and τ is a K-section, i.e., a K-homomorphism $C \longrightarrow B$ such that $\tau\pi = id_C$.

For each g in G and c in C,

$$cg^{\tau} - c^{\tau}g = cg' \in A$$

and $g' : c \mapsto cg'$ is a K-linear map of C into A:

$$g' \in H = \mathrm{Hom}_K(C,A). \qquad (2)$$

If $g_1 \in G$, $(g\, g_1)' = g'g_1 + g\, g_1'$ and so $d : g \mapsto g'$ satisfies

$$(g\, g_1)d = (g\, d)g_1 + g(g_1 d), \qquad (3)$$

where H is regarded as a two-sided KG-module in the usual manner (cf. p.15).

<u>Definition</u>. If V is a two-sided KG-module, then a mapping $d : G \to V$ is called a <u>derivation</u> of G in V if

$$(g\, g_1)d = (g\, d)g_1 + g(g_1 d)$$

for all g, g_1 in G. The set of all these forms a K-module Der(G,V). If $v \in V$, $g \mapsto -vg + gv$ is called an <u>inner derivation</u>; and all these form a K-module Ider(G,V).
(These definitions reduce to the earlier ones in Chapter 3 if the left action of G on V is trivial: then V is simply a right module.)

Thus by (3) above, d is a derivation of G in H. If τ' is a second K-section and d' the corresponding derivation, then $d - d'$ is inner, determined by $\tau - \tau' \in H$.

Suppose $\quad 0 \to A \to B' \to C \to 0 \qquad (1')$

is a second extension. We call (1), (1') equivalent if there exists a KG-isomorphism $B \xrightarrow{\sim} B'$ restricting to the identity on A and on C in the natural way. Write the set of equivalent extensions as Ext(C,A). Thus we have a mapping

$$\mathrm{Ext}(C,A) \longrightarrow \frac{\mathrm{Der}(G,H)}{\mathrm{Ider}(G,H)}.$$

Conversely, let $d \in \mathrm{Der}(G,H)$, $B = A \oplus C$ as K-module and define an action of G on B by

$$(a,c)g = (ag - c(gd),\ cg).$$

Then $0 \to A \to B \to C \to 0$ is KG-exact, where $a \mapsto (a,0)$, $(a,c) \mapsto c$, and the section $c \mapsto (0,c)$ yields d. If $f \in \mathrm{Ider}(G,H)$, $d + f$ determines, by the above construction, an equivalent extension. Hence we have

PROPOSITION 1. *If A is K-injective or C is K-projective and* $H = \mathrm{Hom}_K(C,A)$,

$$\mathrm{Ext}(C,A) \xrightarrow{\sim} \mathrm{Der}(G,H)/\mathrm{Ider}(G,H).$$

The zero corresponds to the split extension.

Let V be any two-sided KG-module. Then V is a right $K(G \times G)$-module by $v(g, g_1) = g_1^{-1}v\, g$, and so V is a right KG-module via $G \to G \times G$, $g \mapsto (g,g)$: $vg = g^{-1}v\, g$. We shall write V_χ for V when viewed thus. (For H as in (2) above, H_χ is H qua right KG-module via conjugation in the usual way: p.15.)

If now $d \mapsto d^{\rightarrow}$: $\mathrm{Der}(G,V) \to \mathrm{Der}(G,V_\chi)$ by

$$g\, d^{\rightarrow} = (g^{-1}(g\, d))_\chi\ ,$$

then it is easily checked that this is a K-isomorphism (the inverse is $d \mapsto d^{\leftarrow}$, where $gd^{\leftarrow} = g(g\, d)$). Moreover, $d^{\rightarrow}$ is inner if, and only if, d is.

Proposition 1 and the formula on p.45 now yields

PROPOSITION 2. *If A is K-injective or C is K-projective,*

$$\mathrm{Ext}(C,\ A) \cong H^1(G,\ \mathrm{Hom}_K(C,\ A)_\chi\ ).$$

COROLLARY. (Maschke's Theorem.)   If |G| is invertible in K, then
K-projective implies KG-projective and K-injective implies
KG-injective.

Proof.   By Corollary 1, p.91, $H^1(G, M) = 0$ for all KG-modules M.
So if C is K-projective, Ext(C,A) = 0 for all A; and if A is
K-injective, Ext(C,A) = 0 for all C.

Remarks.

(1)   : Propositions 1 and 2 are true without the hypothesis
that G is finite.

(2)   : The Corollary shows that K is KG-projective when
|G|K = K.   We already know this from Proposition 21, p.176.

(3)   : Ext(C,A) is effectively the same as $Ext_{KG}^1(C,A)$ and
the isomorphism of Proposition 2 is a special case of the general
formula

$$Ext_{KG}^m(C,A) \simeq H^m(G, Hom_K(C,A)_\lambda).$$

(Cartan - Eilenberg, p.226)

PROPOSITION 3.   Ext(C,A) = 0 if either
        (i) C is K-projective and A is KG-projective;
or   (ii) C is KG-injective and A is K-injective.

        We split the proof into three simple lemmas.   Recall(from
the exercise on p.16) that for any K-module V,

$$Hom_K(KG,V)_\lambda \simeq V^*.$$

We use this repeatedly in what follows.

<u>LEMMA 1. If B is a KG-module and B^0 denotes the K-module B with trivial G-action, then (i) $B \otimes KG \simeq B^0 \otimes KG$ (diagonal action) and (ii) $\mathrm{Hom}_K(KG, B)_\kappa \simeq \mathrm{Hom}_K(KG, B^0)$.</u>

<u>Proof</u>. (i) $bg \otimes g \mapsto b \otimes g$.

(ii) $\mathrm{Hom}_K(KG, B^0) \simeq B^0 \otimes KG$ (cf. p.22)

$\simeq B \otimes KG$ (part (i))

$\simeq B \otimes \mathrm{Hom}_K(KG, K)$ (p.22).

Finally, $\varphi : B \otimes \mathrm{Hom}_K(KG, K) \to \mathrm{Hom}_K(KG, B)_\kappa$ defined by

$$(b \otimes f)\varphi : x \to (x^{-1}f)b \qquad (x \in G)$$

is a KG-isomorphism (the inverse being

$$f \mapsto \sum_{x \in G} xf \otimes x^*,$$

where $y\, x^* = \delta_{yx,\, 1}$).

<u>Exercises</u>.

1. (Generalisation of the isomorphism φ in the proof of Lemma 1.) If $C \in \mathrm{Lat}_{KG}$, then $\mathrm{Hom}_K(C,K) \otimes A \simeq \mathrm{Hom}_K(C,A)_\kappa$. (For an even more general setting cf. Bourbaki, Algèbre, chapter 2 (3rd ed.) §4.2.)

2. If $C \in \mathrm{Lat}_{KG}$ and P is KG-projective, then $C \otimes P$ is KG-projective and deduce that $\mathrm{Hom}_K(C,P)_\kappa$ is also KG-projective. (Use Lemma 1(i) and exercise 1.)

<u>LEMMA 2. For any KG-modules A, B, C,</u>

$$\mathrm{Hom}_K(C, \mathrm{Hom}_K(B,A)_\kappa)_\kappa \simeq \mathrm{Hom}_K(C \otimes B, A)_\kappa .$$

Proof. $f \mapsto (c \otimes b \mapsto b(cf))$.

LEMMA 3. $\mathrm{Hom}_K(C,A^*)_\kappa \cong \mathrm{Hom}_K(C_*,A)_\kappa$ and is coinduced.

Proof. Since $C_* = KG \otimes C^o$ (notation as in Lemma 1),

$$\begin{aligned}\mathrm{Hom}_K(C_*,A)_\kappa &\cong \mathrm{Hom}_K(KG,\ \mathrm{Hom}_K(C^o,A)_\kappa\ )_\kappa && \text{(Lemma 2)}\\ &\cong \mathrm{Hom}_K(C,A)^* && \text{(Lemma 1(ii))}.\end{aligned}$$

In particular, this is true with A^o. But

$$\begin{aligned}\mathrm{Hom}_K(C_*,A^o) &\cong \mathrm{Hom}_K(C \otimes KG,\ A^o) && \text{(Lemma 1(i))}\\ &\cong \mathrm{Hom}_K(C,A^*)_\kappa && \text{(Lemma 2)}.\end{aligned}$$

Proof of Proposition 3. In view of Proposition 2 it is (more than) sufficient to show that $\mathrm{Hom}_K(C,A)_\kappa$ is a direct summand of some coinduced module.

In case (i), A is a direct summand of A_* and $A_* \cong A^*$ (p.22). So we may replace A by A^*. In case (ii), for similar reasons, we may replace C by C_*. The result now follows from Lemma 3.

COROLLARY. If K is a field, KG-projective = KG-injective.

Proof. B is KG-projective $\Rightarrow$ Ext(C,B) = 0 for all C (Proposition 3 (i)) $\Rightarrow$ B is KG-injective (by embedding B in an injective module) $\Rightarrow$ Ext(B,A) = 0 for all A (Proposition 3(ii)) $\Rightarrow$ B is KG-projective (by presenting B as the image of a projective).

Exercise. If K is such that K-projective = K-injective, prove that KG-projective = KG-injective.

10.2   Heller modules.

Definition.   If $C \in \text{Lat}_{KG}$, we shall call C a <u>Heller module</u> if $C \neq 0$, C is not KG-projective and

(*) $C \oplus P = A \oplus B$, with P projective, implies A or B projective.

PROPOSITION 4.   Assume KG has the following property: Every finitely generated KG-module A admits a decomposition $A = A' \oplus A''$, where A'' is KG-projective, A' has no KG-projective direct summand and A' is unique to within isomorphism: i.e., $A' \oplus A'' \simeq B' \oplus B''$ implies $A' \simeq B'$.

Let $C \in \text{Lat}_{KG}$.   Then C is a Heller module if, and only if, C' is non-zero and indecomposable.

A sufficient condition for KG to have the property of this proposition is that K be noetherian (this ensures that every A decomposes as required) and that the usual Krull-Schmidt theorem holds for KG (this ensures the uniqueness of A': it also yields the uniqueness of A'', but this is of no interest in our context.)

So all is certainly well if K is a field.   More generally, we quote here a companion to the Projective Cover Theorem of §9.8.

KRULL-SCHMIDT THEOREM.   Let I be an ideal of the commutative noetherian ring K so that K is I-complete and K/I satisfies the decending chain condition on ideals.   If G is finite, then every finitely generated KG-module decomposes into a direct sum of indecomposable modules and this decomposition is unique to within order and isomorphism of the factors.

Theorem 2.20 of [4], p.87, provides a proof provided one also uses the following fact: with K as above, any finitely generated K-algebra A is noetherian, J-complete where J is the Jacobson radical of A, and A/J has the descending chain condition on right ideals: cf. the proof of Corollary 2.22 in [4], p.88.

The Krull-Schmidt theorem is not, however, necessary for Proposition 4 to work. This will be of importance to us later: cf. §10.6 and §11.7.

Proof of Proposition 4. If $C' = U \oplus V$, then U, V are non-projective and hence the Heller condition (*) fails (with $P = 0$). So C is not Heller.

Conversely, assume $C' \neq 0$ and is indecomposable. Let $C \oplus P = A \oplus B$. By our hypothesis on KG, $C' \cong A' \oplus B'$ and hence A' or B' is 0 since C is indecomposable. Thus A or B is projective.

PROPOSITION 5. (Heller [2].) Let $0 \longrightarrow A \longrightarrow P \longrightarrow C \longrightarrow 0$ be exact in Lat_{KG}, with P KG-projective. If C is a Heller module, then so is A.

If C is a KG-module, we shall write $C^* = \mathrm{Hom}_K(C,K)$ (action: $\varphi g : c \longmapsto cg^{-1}\varphi$), called the dual (K-dual) of C. If C is in Lat_{KG}, so is C^*.

LEMMA 4. (i) If $C \in \mathrm{Lat}_{KG}$, the natural mapping $C \longrightarrow C^{**}$ is a KG-isomorphism.

(ii) If C is KG-projective, so is C^*.

(iii) If $C \in \mathrm{Lat}_{KG}$, then C is a Heller module if, and only if, C^* is one.

Proof. (i): exercise;

(ii): $KG \cong (KG)^*$ (coinduced);

(iii): By (i) and (ii), C is non-projective if, and only if, C^* is.

Assume C is a Heller module and let $C^* \oplus P = A \oplus B$. Then $C \oplus P^* = A^* \oplus B^*$, using (i) and so A^*, say, is projective. Therefore $A \cong A^{**}$ is projective by (i) and (ii), so that C^* is Heller.

Proof of Proposition 5. As C is not projective, $A \neq 0$ and A is non-projective (Proposition 3(i)). Suppose

$$A \oplus P_o = A_1 \oplus A_2,$$

where P_o is projective but A_1, A_2 are not projective. Let

$$0 \longrightarrow U_i \longrightarrow V_i \longrightarrow A_i^* \longrightarrow 0, \qquad i = 1,2,$$

be exact, with V_1, V_2 projective. Then U_1, U_2 are non-projective (Proposition 3(i)).

Now Schanuel (Lemma 11, §8.10, p.162) applied to

$$0 \longrightarrow C^* \longrightarrow P^* \oplus P_o^* \longrightarrow A^* \oplus P_o^* \longrightarrow 0,$$

$$0 \longrightarrow U_1 \oplus U_2 \longrightarrow V_1 \oplus V_2 \longrightarrow A_1^* \oplus A_2^* \longrightarrow 0$$

yields

$$C' \oplus V_1 \oplus V_2 \cong (U_1 \oplus P') \oplus (U_2 \oplus P_o').$$

By Lemma 4, C' is a Heller module. Hence $U_1 \oplus P'$ or $U_2 \oplus P_o'$ is projective and so U_1 or U_2 is projective (Lemma 4, (ii)).

This contradiction shows that A satisfies (*) and hence A is a Heller module.

Problem. Is $\mathbb{Z}$ a Heller module?

10.3 Ext under flat coefficient extensions.

Let $K \to S$ be a homomorphism of commutative rings and assume S as a K-module is K-flat: this means that $M \mapsto M \otimes_K S = M_{(S)}$ is an exact functor from Mod_K to Mod_S.

For any KG-modules C, A we define

$$\omega : \quad Hom_K(C, A)_{(S)} \to Hom_S(C_{(S)}, A_{(S)})$$

by $(f \otimes s)\omega : c \otimes s' \mapsto cf \otimes ss'$ $(f \in Hom_K(C, A), s \in S)$. This is a KG-homomorphism. Since $(M^G)_{(S)} \leq (M_{(S)})^G$ for any KG-module M, ω induces a homomorphism

$$Hom_{KG}(C, A)_{(S)} \to Hom_{SG}(C_{(S)}, A_{(S)}).$$

Given a finitely generated C, we may always find an epimorphism $\varphi : E \to C$ with E KG-free and finitely generated. But $Ker\varphi$ may not be finitely generated.

<u>Definition</u>. If φ exists with $\mathrm{Ker}\varphi$ finitely generated, we call C <u>finitely presentable</u> (as KG-module).

Note that, if K is noetherian, every finitely generated module is finitely presentable.

<u>Exercises</u>.

1. Given exact sequences of K-modules

$$0 \longrightarrow A \longrightarrow E \longrightarrow C \longrightarrow 0,$$

$$0 \longrightarrow A' \longrightarrow E' \longrightarrow C \longrightarrow 0,$$

where E is K-free and E', C, A are finitely generated, then A' is finitely generated. (Hint: Lift $E \longrightarrow C$ to ψ: $E \longrightarrow E'$ and let the restriction of ψ to A be θ. Show that ψ, θ have isomorphic cokernels.)

2. C is finitely presentable as KG-module if, and only if, C is finitely presentable as K-module. (Use exercise 1.)

<u>PROPOSITION 6. If C is finitely presentable, ω (as defined above) is an isomorphism and induces an isomorphism</u>

$$\underline{\mathrm{Hom}_{KG}(C, A)_{(S)} \xrightarrow{\sim} \mathrm{Hom}_{SG}(C_{(S)}, A_{(S)})}.$$

<u>Proof</u>. Let C be K-free on $c_1, \ldots, c_n$ and choose any $\varphi \in \mathrm{Hom}_S(C_{(S)}, A_{(S)})$. If

$$(c_i \otimes 1)\varphi = \sum_j a_{ij} \otimes s_{ij} \qquad (a_{ij} \in A,\ s_{ij} \in S),$$

then define $\eta_{ij} : C \longrightarrow A$ by

$$c_k \eta_{ij} = \delta_{ki}\, a_{ij}$$

and set

$$\varphi^+ = \sum_{i,j} \eta_{ij} \otimes s_{ij} \in \mathrm{Hom}_K(C,\ A)_{(S)}\ .$$

Clearly $\varphi \mapsto \varphi^+$ is the inverse of ω.

Now suppose $0 \to B \to E \to C \to 0$ is exact with B finitely generated and E KG-free. Since S is K-flat, the rows of the following diagram are exact:

$$\begin{array}{ccccc} 0 \to \mathrm{Hom}_K(C,A)_{(S)} & \longrightarrow & \mathrm{Hom}_K(E,A)_{(S)} & \longrightarrow & \mathrm{Hom}_K(B,A)_{(S)} \\ \downarrow \omega & & \downarrow \omega & & \downarrow \omega \\ 0 \to \mathrm{Hom}_S(C_{(S)},A_{(S)}) & \to & \mathrm{Hom}_S(E_{(S)},A_{(S)}) & \to & \mathrm{Hom}_S(B_{(S)},A_{(S)}) \end{array} .$$

The middle ω is an isomorphism (just proved above) and therefore the left hand ω is injective. Thus ω is injective for <u>all</u> finitely generated modules C. In particular this holds for B and hence the right hand ω is injective. Consequently the left ω is an isomorphism.

The last part of the proposition is a consequence of

<u>LEMMA 5. If M is a KG-module, $(M^G)_{(S)} = (M_{(S)})^G$.</u>

<u>Proof</u>. (Fröhlich) Let $M_o = \bigoplus_{g \in G} M_g$, where $M_g \cong M$ and

$$\psi\colon M \to M_o \quad \text{be} \quad a \mapsto (a(g-1))_g\ .$$

Then ψ is K-linear and $\mathrm{Ker}\,\psi = M^G$. Now

$$0 \to (M^G)_{(S)} \to M_{(S)} \xrightarrow{\psi_{(S)}} (M_o)_{(S)}$$

is exact (as S is K-flat) and $\psi_{(S)}$ has kernel $(M_{(S)})^G$.

PROPOSITION 7. For all $k \geq 0$,

$$H^k(G, A)_{(S)} \cong H^k(G, A_{(S)}).$$

Proof. Since K is finitely presentable, the result is true for $k = 0$ by Proposition 6 ($H^0(G,A) = \mathrm{Hom}_{KG}(K,A)$).

Let $\ldots \to P_1 \to P_0 \to K \to 0$ be a KG-projective resolution of K with each P_i finitely generated (and therefore finitely presentable!). If $Y = \mathrm{Im}(P_k \to P_{k-1})$, we have the diagram

$$\begin{array}{ccccccc} \mathrm{Hom}_{KG}(P_{k-1},A)_{(S)} & \longrightarrow & \mathrm{Hom}_{KG}(Y,A)_{(S)} & \longrightarrow & H^k(G,A)_{(S)} & \longrightarrow & 0 \\ \downarrow{\scriptstyle\omega} & & \downarrow{\scriptstyle\omega} & & & & \\ \mathrm{Hom}_{SG}((P_{k-1})_{(S)},A_{(S)}) & \to & \mathrm{Hom}_{SG}(Y_{(S)},A_{(S)}) & \to & H^k(G,A_{(S)}) & \to & 0, \end{array}$$

where the rows are exact and the vertical down maps are isomorphisms by Proposition 6. These ω's induce the required isomorphism.

COROLLARY. If $C \in \mathrm{Lat}_{KG}$, $\mathrm{Ext}(C, A)_{(S)} \cong \mathrm{Ext}(C_{(S)}, A_{(S)})$.

This is immediate by Propositions 2, 6 and 7.

Remark. Proposition 7 is a special case of the following: for all KG-modules C that admit a finitely generated KG-projective resolution and all $k \geq 0$,

$$\mathrm{Ext}_{KG}{}^k(C,A)_{(S)} \cong \mathrm{Ext}_{SG}{}^k(C_{(S)},A_{(S)}).$$

10.4 Localisation.

We recall some standard material.

Let K be an integral domain and let Z be a multiplicatively closed subset of K containing 1 but not containing 0. We write the resulting subring of the quotient field of K,

$$\{a/b; \quad a \in K, \ b \in Z\}$$

as $Z^{-1}K$. If A is a K-module, put

$$Z^{-1}A = A \otimes_K Z^{-1}K.$$

$Z^{-1}K$ is always K-flat and hence <u>all the results of the previous section apply with $Z^{-1}K$ in place of S.</u>

Elementary facts:

E1. $\mathrm{Ker}(A \longrightarrow Z^{-1}A) = \{a \in A; \quad za = 0 \text{ for some } z \text{ in } Z\}$;

E2. if $Z \leq Y$, $(Z^{-1}Y)^{-1}(Z^{-1}K) = Y^{-1}K$;

E3. if K is noetherian, so is $Z^{-1}K$.

PROPOSITION 8. <u>Let K be a noetherian domain. Then</u>

$$Z^{-1}(\ ): \ \mathrm{fMod}_{KG} \longrightarrow \mathrm{fMod}_{Z^{-1}KG}$$

<u>is a surjective functor (cf. p.187).</u>

(Here fMod_* means the full subcategory of all finitely generated modules over *.)

<u>Proof</u>. Let $L \in \mathrm{fMod}_{Z^{-1}KG}$ and suppose

$$(Z^{-1}KG)^m \xrightarrow{\varphi} (Z^{-1}KG)^n \longrightarrow L \longrightarrow 0$$

is exact. (L is the image of a finitely generated free module and the kernel is finitely generated since $Z^{-1}KG$ is noetherian.)

By Proposition 6, we can find $f \in \mathrm{Hom}_{KG}((KG)^m,(KG)^n)$ and $r \in Z$ so that $r\varphi = f \otimes 1$.

Since multiplication by r is an automorphism of $Z^{-1}K$-modules,

$$L \cong \operatorname{coker}\varphi = \operatorname{coker} r\varphi = \operatorname{coker}(f \otimes 1) \cong Z^{-1}(\operatorname{coker} f).$$

This proves the surjectivity of $Z^{-1}(\ )$ on objects. The surjectivity on morphism sets is a consequence of Proposition 6.

Note that, without the noetherian hypothesis on K in Proposition 8, the argument shows that if L is a finitely <u>presentable</u> $Z^{-1}KG$-module, then $L \cong Z^{-1}M$, where M is a finitely presentable KG-module.

Suppose $\mathfrak{p}$ is a prime ideal of K. Then $Z = K - \mathfrak{p}$ is multiplicatively closed and contains 1 but not 0. One usually writes $A_{\mathfrak{p}}$ instead of $Z^{-1}A$. In particular, $K_{\mathfrak{p}} = Z^{-1}K$ and in this ring every element outside the ideal $Z^{-1}\mathfrak{p} = \{a/b;\ a \in \mathfrak{p},\ b \notin \mathfrak{p}\}$ is invertible.

<u>Definition</u>. A ring in which all the non-invertible elements form an ideal $\mathfrak{m}$ is called a <u>local ring</u>. Then $\mathfrak{m}$ is necessarily the unique maximal ideal.

<u>Exercise</u>. L is local if, and only if, L/Jac(L) is a division ring. (Jac(L) = Jacobson radical of L.)

Thus $K_{\mathfrak{p}}$, above, is a local ring. Important example: $\mathbb{Z}_{(p)}$, for p a prime number.

Consider now a Dedekind domain K. This means that K is an integral domain in which every ideal is uniquely expressible as a product of prime ideals. (For the basic facts concerning Dedekind domains see, e.g., [6], chapter 5.)

Assume that A is a torsion module over K. If $\mathfrak{p}$ is prime (= maximal) in K, let

$$A(\mathfrak{p}) = \{a \in A;\ a\mathfrak{p}^s = 0 \text{ for some } s\}.$$

Then we have the following further elementary facts:

E4. $A = \coprod_{\mathfrak{p}} A(\mathfrak{p})$;

E5. $A(\mathfrak{p}) \cong A_{\mathfrak{p}}$.

An immediate consequence of this, Proposition 7 and Corollary 1 on p.91 is

PROPOSITION 9. Let K be a Dedekind domain in which $|G| \neq 0$. Then for all $k > 0$,

$$H^k(G, A) \cong \coprod_{|G| \in \mathfrak{p}} H^k(G, A_{\mathfrak{p}}).$$

COROLLARY 1. If $C \in \mathrm{Lat}_{KG}$,

$$\mathrm{Ext}(C,A) \cong \coprod_{|G| \in \mathfrak{p}} \mathrm{Ext}(C_{\mathfrak{p}}, A_{\mathfrak{p}}).$$

Proof. Apply Propositions 2 and 6.

Let $Z = \bigcap_{|G| \in \mathfrak{p}} (K - \mathfrak{p})$. Then Z is multiplicatively closed,

$1 \in Z$ and $0 \notin Z$. We shall write $Z^{-1}K = K_{(G)}$ and $Z^{-1}A = A_{(G)}$.

Note that $\mathbb{Z}_{(G)}$ consists of all rational numbers a/b where b is prime to $|G|$.

<u>Exercises</u>. (Cf. Bourbaki, Algèbre Commutative, chapter 2, §3, nos. 3,5.)

1. If I is an integral domain and Ω is the set of all maximal ideals show that $I = \bigcap_{\mathfrak{m} \in \Omega} I_{\mathfrak{m}}$.
(If $x \in$ r.h.s., let $\mathfrak{a} = \{a \in I \mid ax \in I\}$. Then $x \notin$ l.h.s. if, and only if, $\mathfrak{a} \neq I$.)

2. Let I be an integral domain and Z a multiplicatively closed subset containing 1 but not 0. If Ω is all primes $\mathfrak{p}$ in I such that $\mathfrak{p} \cap Z = \phi$ prove that $\mathfrak{p} \mapsto Z^{-1}\mathfrak{p}$ is a one-one mapping of Ω onto the set of all prime ideals of $Z^{-1}I$.

3. If Ω is a finite set of primes in the Dedekind domain K and $Z = \bigcap_{\mathfrak{p} \in \Omega} (K - \mathfrak{p})$, prove that $\mathfrak{q} \in \Omega \cup \{0\}$ if, and only if, $\mathfrak{q} \cap Z = \phi$.
(If a prime lies in the union of a finite number of primes, then it lies in one of them.)

4. If K, Ω, Z are as in exercise 3, show that

$$Z^{-1}K = \bigcap_{\mathfrak{p} \in \Omega} K_{\mathfrak{p}} .$$

In particular, $K_{(G)} = \bigcap_{|G| \in \mathfrak{p}} K_{\mathfrak{p}}$.

(Use exercises 1, 2, 3, and "elementary fact" E2 above.)

COROLLARY 2. (i) For all $k > 0$,

$$H^k(G,A) \cong H^k(G,A_{(G)}).$$

(ii) If $C \in \mathrm{Lat}_{KG}$,

$$\mathrm{Ext}(C,A) \cong \mathrm{Ext}(C_{(G)}, A_{(G)}).$$

Proof. Use the elementary fact E2 above with $Z = \bigcap_{|G| \in \mathfrak{p}} (K - \mathfrak{p})$ and $Y = K - \mathfrak{p}$; and apply Proposition 9 for part (i), and Corollary 1 for part (ii), to KG and $K_{(G)}G$.

COROLLARY 3. Let $C \in \mathrm{Lat}_{KG}$. Then C is KG-projective if, and only if, $C_{(G)}$ is $K_{(G)}G$-projective.

Proof. If $C_{(G)}$ is $K_{(G)}G$-projective, then the right hand side of the isomorphism of Corollary 2 vanishes for all A and hence so does the left side. Thus C is KG-projective.

Conversely, if $C \oplus A$ is KG-free, then $C_{(G)} \oplus A_{(G)}$ is $K_{(G)}G$-free.

Exercise. Show that the condition $|G| \neq 0$ in Proposition 9 is necessary. (Consider G cyclic of order p and $A = K = \mathbb{F}_p[X]$.)

10.5 Local rings.

If the coefficient ring K is local, this does not in general imply the same of KG: For example, if K is any field of characteristic prime to $|G|$, KG is semi-simple and so, if

KG were local, KG would be a division ring, whence $\mathfrak{g}$ (the augmentation ideal) would be zero, and thus $G = 1$.

PROPOSITION 10. KG is local if, and only if, K is local and G is a p-group where p is the characteristic of $K/\mathfrak{m}$, the residue class field of K.

Proof. If KG is local, so is every homomorphic image. In particular, $K \cong KG/\mathfrak{g}$, is local, with maximal ideal $\mathfrak{m}$ say. Let $F = K/\mathfrak{m}$ and p = characteristic of F. Now $KG/\mathfrak{m}G \cong FG$ is also local and hence $p \neq 0$ and the augmentation ideal of FG is nilpotent. Thus $g-1$ is nilpotent (in FG) for all g and so $g^{p^r} = 1$ for some r.

Conversely, assume G is a p-group and K a local ring with $F = K/\mathfrak{m}$ of characteristic p. We shall prove $KG/\mathrm{Jac}(KG) \cong F$. Observe first that

$$\mathrm{Jac}(KG) \geq \mathfrak{m}G.$$

For if M is a simple KG-module, $M\mathfrak{m} = 0$ by Example 4, p.97. Hence we need only show that FG is local. Now for each g in G, $g^{p^r} = 1$ for some r and so $g-1$ is nilpotent. Hence the augmentation ideal $\mathfrak{g}$ of FG is F-spanned by nilpotent elements and therefore is nilpotent by a theorem of Wedderburn [5]. Consequently $\mathfrak{g} = \mathrm{Jac}(FG)$, as needed.

We record here a consequence of Proposition 10 that we shall need later (Theorem 4, §10.7).

<u>PROPOSITION 11. Let K be a field of characteristic p and G a p-group. If the KG-module A satisfies $H_1(G,A) = 0$, then A is KG-free.</u>

<u>Proof</u>. Let E be a free KG-module on a basis (e_i) in one-one correspondence with a K-basis $(a_i + A\mathfrak{g})$ of $A/A\mathfrak{g}$. Then $e_i \mapsto a_i$ yields a homomorphism $\varphi : E \to A$ so that $\mathrm{Im}\varphi + A\mathfrak{g} = A$. Hence φ is surjective because $\mathfrak{g}$ is nilpotent. The exact homology sequence associated with

$$0 \to T \to E \xrightarrow{\varphi} A \to 0$$

shows that

$$0 = H_1(G,A) \to H_0(G,T) \to H_0(G,E) \to H_0(G,A) \to 0$$

is exact, whence $H_0(G,T) = T/T\mathfrak{g} = 0$. So $T = 0$.

10.6 Semi-local coefficients.

<u>Definition</u>. A ring L is called <u>semi-local</u> if L has only a finite number of maximal ideals.

<u>Exercises</u>.

1. L is semi-local if, and only if, L/Jac(L) has descending chain condition on right ideals (i.e., is semi-simple).
2. The ring $K_{(G)}$ introduced above immediately after Corollary 1 to Proposition 9 (§10.4, p.236) is semi-local.
3. If K is semi-local, so is KG. (Use $\mathrm{Jac}(K)G \leq \mathrm{Jac}(KG)$.)

Group algebras over semi-local rings enjoy two important properties that we shall need later.

<u>THEOREM 1. (The Cancellation Theorem). If K is semi-local,</u>

$$A \oplus KG \cong B \oplus KG$$

<u>implies $A \cong B$.</u>

The proof is very accessible in Swan [4], p.176; or Bass [1], p.168.

<u>THEOREM 2. Let K be a semi-local Dedekind domain of characteristic zero in which every prime dividing $|G|$ is non-invertible. Then every finitely generated projective KG-module is KG-free.</u>

This theorem is an immediate consequence of the following marvellous result.

<u>SWAN'S THEOREM [3]. Let K be a Dedekind domain of characteristic zero in which every prime dividing $|G|$ is non-invertible. Then every finitely generated projective KG-module is of the form $I \oplus (KG)^r$ for some ideal I in KG. Moreover, if $\mathfrak{a}$ is any non-zero ideal in K, I can be chosen so that $I \cap K$ is prime to $\mathfrak{a}$.</u>

If K in this theorem is semi-local, then $\mathrm{Jac}(K) \neq 0$. So we may assume $I \cap K$ is prime to Jac(K). But this implies $I \cap K = K$ and so $I = KG$. Thus we have Theorem 2.

A consequence of Theorems 1 and 2 is the following

COROLLARY. Let K be a semi-local Dedekind domain of characteristic zero in which every prime dividing $|G|$ is non-invertible. Then KG has the property required in Proposition 4 of §10.2 (p.227).

Proof. Since K is noetherian, every finitely generated KG-module decomposes as required. Suppose $A' \oplus A'' \cong B' \oplus B''$, with A'', B'' projective but no projective direct summand in A' or B'. Then A'', B'' are free (Theorem 2), say of ranks r, r+s, respectively, and thus (Theorem 1), $A' \cong B' \oplus (KG)^s$. Consequently s = 0 by our hypothesis on A'.

10.7 Cohomological criteria for projectivity.

THEOREM 3. Let G be a group and K an integral domain such that, for every p dividing $|G|$, either pK = K or pK is a maximal ideal. If $A \in \mathrm{Lat}_{KG}$ and satisfies

$$H^m(G_p, A) = H^{m+1}(G_p, A) = 0$$

for some $m \geq 1$ and all $p/|G|$ (where G_p is any Sylow p-subgroup of G), then A is KG-projective.

Theorem 3 can be applied when K is $\mathbb{Z}$, $\mathbb{Z}_{(G)}$, $\mathbb{Z}_{(p)}$, or $\mathbb{Z}_p$, the p-adic integers. It can also of course be applied when K is a field. But in that case a stronger result holds.

THEOREM 4. Suppose F is a field of characteristic p and $H^m(G_p, A) = 0$ for some $m \geq 1$. Then

(i) A is FG-projective and

(ii) A is FG_p-free.

Part (i) is an immediate consequence of part (ii) and Theorem 3 (because $H^k(G_q, A) = 0$ for all $k \geq 1$ and all $q \neq p$ by Maschke's theorem (p. 224)).

Proof of Theorem 3 (assuming part (ii) of Theorem 4). Choose

$$0 \to B \to U \to A \to 0$$

exact with U KG-free and finitely generated. Since A is K-projective,

$$0 \to \mathrm{Hom}_K(A,B) \to \mathrm{Hom}_K(A,U) \to \mathrm{Hom}_K(A,A) \to 0 \qquad (1)$$

is still exact. Now A is KG-projective if, and only if, U splits over B, i.e., if and only if

$$\mathrm{Hom}_{KG}(A,U) \to \mathrm{Hom}_{KG}(A,A)$$

is surjective; and this holds if, and only if

$$H^1(G, \mathrm{Hom}_K(A,B)) = 0 \qquad (2)$$

(use the cohomology sequence corresponding to (1)). Writing $M = \mathrm{Hom}_K(A,B)$, we see that by Corollary 2, p.91, (2) is true if, and only if,

$$H^1(G_p, M) = 0, \text{ for all } p. \qquad (3)$$

Since A, B are both in Lat_{KG}, so is M. Hence if $pK = K$, M is KG_p-projective by Maschke's theorem (p.224). It therefore remains to prove (3) for all those p such that $pK \neq K$. Choose one such p and set $F = K/pK$, which is a field by our hypothesis.

We assert that (3) holds for p if

$$H^1(G_p, M/Mp) = 0: \tag{4}$$

When p is the characteristic of K, (4) coincides with (3). If $p \neq 0$ in K, the sequence $0 \to M \xrightarrow{p} M \to M/Mp \to 0$ is exact because K is an integral domain and M is K-projective; and it yields, assuming (4),

$$H^1(G_p, M) = p\, H^1(G_p, M).$$

But $|G_p|$ annihilates $H^1(G_p, M)$ (Corollary 1, p.91) and so $H^1(G_p, M) = 0$, as needed.

If $p \neq 0$ in K, $0 \to B \xrightarrow{p} B \to B/Bp \to 0$ is exact (as B is K-projective). So we obtain the exact sequence

$$0 \to M \xrightarrow{p} M \to \mathrm{Hom}_K(A, B/Bp) \to 0$$

(as A is K-projective). Hence

$$M/Mp \cong \mathrm{Hom}_F(A/Ap, B/Bp) = V, \text{ say.}$$

If $p = 0$ in K, $M = V$. So in any case we must prove

$$H^1(G_p, V) = 0. \tag{5}$$

Again assume $p \neq 0$ in K. Then our hypothesis on A applied to the cohomology sequence coming from $0 \to A \xrightarrow{p} A \to A/Ap \to 0$ yields $H^m(G_p, A/Ap) = 0$. When $p = 0$ in K we still have $H^m(G_p, A) = 0$. So A is FG_p-free by Theorem 4 (ii), whence V is a direct sum of modules $\mathrm{Hom}_F(FG_p, B/Bp)$ and these are coinduced (Lemma 1 (ii), p. 225). Thus (5) holds.

We now begin the proof of Theorem 4 (ii), with two lemmas in which G is an arbitrary (finite) group and K is quite unrestricted. As usual, $\tau = \sum_{x \in G} x$, $\mathfrak{g}$ is the augmentation ideal of KG and, for a KG-module A, $A_\tau = \{a \in A \mid a\tau = 0\}$.

<u>LEMMA 6. If $A^G = A\tau$ and $B = \mathrm{Coker}(A \to A^*)$, then $B_\tau = B\mathfrak{g}$.</u>

<u>Proof</u>. Observe first that $(A^*)_\tau = A^*\mathfrak{g}$: for $A^* \cong A_*$ (p.22) and $(\sum_{x\in G} u_x \otimes x)\tau = 0$ implies $\Sigma u_x = 0$ so that

$$\Sigma u_x \otimes x = \Sigma(u_x \otimes 1)(x-1).$$

Let b in B satisfy $b\tau = 0$. If b is the image of some f in A^*, $f\tau = a^*$, for a suitable a in A. Then $a \in A^G$, whence $a = a_1\tau$ by hypothesis. Hence $(f - a_1^*) \in (A^*)_\tau = A^*\mathfrak{g}$ and so $b \in B\mathfrak{g}$, as required.

<u>LEMMA 7. If $A_\tau = A\mathfrak{g}$ and $B = \mathrm{Coker}(A \to A^*)$, then $H_1(G,B) = 0$.</u>

<u>Proof</u>. The homology sequence for $0 \to A \to A^* \to B \to 0$ gives

$$0 = H_1(G,A^*) \to H_1(G,B) \to A_G \to (A^*)_G \to B_G \to 0.$$

We assert $A_G \to (A^*)_G$ is injective: for if $a^* \in A^*\mathfrak{g} = (A^*)_\tau$, then $a\tau = 0$ and so $a \in A\mathfrak{g}$, by hypothesis.

<u>Remark</u>. It was shown by Tate that the cohomology and homology sequences corresponding to any short exact sequence $0 \to A \to B \to C \to 0$ can be linked to produce a doubly infinite exact sequence

$$\cdots \to H_1(B) \to H_1(C) \to \hat{H}^{-1}(A) \to \hat{H}^{-1}(B) \to \hat{H}^{-1}(C) \to$$
$$\to \hat{H}^{0}(A) \to \hat{H}^{0}(B) \to \hat{H}^{0}(C) \to H^1(A) \to H^1(B) \to \cdots,$$

where $\hat{H}^0(A) = A^G/A\tau$ and $\hat{H}^{-1}(A) = A_\tau/A\mathfrak{g}$.

If we write

$$\hat{H}^r(G,\ ) = H^r(G,\ ) \qquad \text{for } r \geq 1,$$
$$= H_{-(r+1)}(G,\ ) \qquad \text{for } r \leq -2,$$

then one may prove that $(\hat{H}^r(G,\ );\ r \in \mathbb{Z})$ is a connected and minimal sequence of functors, the <u>Tate cohomology</u> of G.

Lemmas 6, and 7, as also Proposition 4, p.95, are immediate consequences of dimension shifting in Tate cohomology. This is merely one illustration of the great advantage of using Tate (as opposed to ordinary) cohomology within finite group theory. For details see Cartan-Eilenberg, chapter 12 and Serre, Corps Locaux, chapter 8.

<u>Proof of Theorem 4(ii)</u>. We shall write G instead of G_p.

Consider the short exact sequences

$$0 \to A_i \to (A_i)^* \to A_{i+1} \to 0 \quad \text{for } i \geq 0,$$

where $A_o = A$. Note that each $A_i{}^*$ is actually FG-free by Proposition 11 (p.240). Now

$$0 = H^m(G,A) \cong H^{m-1}(G,A_1) \cong \dots \cong H^1(G,A_{m-1}).$$

Hence $(A_m)^G = A_m{}^{\tau}$ by Proposition 4, p.95; and then $H_1(G,A_{m+2}) = 0$ (Lemmas 6,7). Thus A_{m+2} is FG-free (Proposition 11,p.240). We conclude that $A_{m+1}, \dots, A_1, A_o$ are all, in turn, FG-free.

Sources and references.

The work of Heller (§10.2) was drawn to my attention by J.A. Green. The material in §10.7 is due to Nakayama and Rim and our account is based on Serre, Corps Locaux, chapter 9.

[1] Bass, H.: Algebraic K-theory, Benjamin, New York 1968.

[2] Heller, A.: Indecomposable representations and the loop-space operation, Proc.Amer.Math.Soc. 12 (1961) 640-643.

[3] Swan, R.G.: Induced representations and projective modules, Annals of Math. 71 (1960) 552-578.

[4] Swan, R.G.: Algebraic K-theory, Springer lecture notes 76, 1968.

[5] Wedderburn, J.H.M.: Note on algebras, Annals of Math. 38 (1937) 854-865.

[6] Zariski, O. and Samuel, P.: Commutative algebra I, van Nostrand, 1958.

A very good reference for commutative algebra (more elementary and, of course, much less complete than Bourbaki) is Introduction to Commutative Algebra by M.F. Atiyah and I.G. Macdonald, Addison-Wesley (1969).

CHAPTER 11

EXTENSION CATEGORIES : FINITE GROUPS

Throughout this chapter the following assumptions will remain in force:

All groups G for which we consider extension categories $\left(\underline{KG}\right)$ *are to be finite groups.*

All KG-modules are to be finitely generated.

We shall use Mod_{KG}, $\left(\underline{KG}\right)$, $\mathcal{Q}_{KG}$ to denote the categories of Chapter 9 but constructed now from the *finitely generated* modules only. (These are therefore full subcategories of the previous categories.)

Remark and warning. Epimorphisms are exactly the surjective morphisms (as before); and enough projective objects exist in all three categories (cf. Remark 2 at the end of §9.7, p.206). But things could go wrong with the monomorphisms and the existence of injectives. (Why? Cf. the first exercise in §9.9, p.211.) However, injectivity questions will not concern us at all in this chapter.

Exercises

1. Given a pair (B,y) with B not necessarily finitely generated.

Prove that there exists a finitely generated submodule B_o and $y_o \in H^2(G,B_o)$ so that inclusion is a morphism: $(B_o,y_o) \hookrightarrow (B,y)$.

2. Denote (here only) the full subcategory of $\mathfrak{Q}_{KG}$ consisting of the pairs (A,x) with A finitely generated by $\mathfrak{Q}_{fKG}$. Prove that (A,x) is essential in $\mathfrak{Q}_{fKG}$ if, and only if, (A,x) is essential in $\mathfrak{Q}_{KG}$. (Use exercise 1.) Deduce that Lemma 5 of §9.8 (p.208) holds in $\mathfrak{Q}_{fKG}$.

Our main concern in this chapter is the study of minimal projectives in $\mathfrak{Q}_{KG}$ for various reasonable choices of the coefficient ring K.

Notation. $\mathbb{F}_p$ is the field of p elements;
$\mathbb{Z}_{(p)}$ is the local ring at p;
$\mathbb{Z}_p$ is the ring of p-adic integers;
$\mathbb{Z}_{(G)} = \cap(\mathbb{Z}_{(p)};\ \text{all } p/|G|) = \{a/b;\ a,b \in \mathbb{Z} \text{ and } (b,|G|)=1\}$.

11.1 Minimal projectives when $|G|$ is invertible in K.

The simplest situation arises if $|G|$ is invertible in K (i.e., $|G|K = K$). If (A,x) is projective in $\mathfrak{Q}_{KG}$, then A is K-projective and hence, in our present case, A is KG-projective (Maschke's Theorem, p.224). Hence $x = 0$. Conversely, if A is KG-projective and $\varphi:\bar{R}_{(K)} \to A$ is an epimorphism, then $\varphi : (\bar{R}_{(K)},0) \to (A,0)$ is a split epimorphism in $\mathfrak{Q}_{KG}$. Thus

(A,0) is projective. This has proved

PROPOSITION 1. Assume $|G|$ is invertible in K. Then

(i) (A,x) is projective in $\mathfrak{Q}_{KG}$ if, and only if, A is KG-projective (and x = 0);

(ii) (A,x) is minimal projective in $\mathfrak{Q}_{KG}$ if, and only if, A is KG-projective and indecomposable.

Thus in the case of a field K of characteristic prime to $|G|$, the theory of the minimal projectives is simply the theory of the irreducible KG-modules, i.e., classical representation theory. In this case all modules are projective. Note however, that Proposition 1 covers situations where the group algebra is not semi-simple (i.e., not all modules are projective): e.g., $K = \mathbb{Z}_{(p)}$, where $p \nmid |G|$.

11.2 Existence of projective covers.

Now suppose $|G|$ is not invertible in K: $|G|K \neq K$. If (A,x) is any projective pair, then x generates $H^2(G,A)$ and $H^2(G,A) \cong K/|G|K$, by Proposition 4, below (§11.3). Hence $x \neq 0$.

Suppose (A,x) is minimal projective but not essential. Then there exists an epimorphism α: $(A,x) \to (C,0)$ with $C \neq 0$. Assume C has a projective cover in Mod_{KG} (§9.8 p.207): $f: P \to C$. Then the projectivity of (A,x) yields a morphism $\varphi:(A,x) \to (P,0)$

so that $\varphi f = \alpha$. Since φ is surjective (projective cover property), P is a direct summand of A, contradicting the minimality of (A,x) (Corollary to Proposition 5, §9.6, p.203).

Hence we have proved

THEOREM 1. *If $|G|K \neq K$ and all KG-modules have projective covers, then $\mathfrak{Q}_{KG}$ contains a projective cover.*

Hence a pair in $\mathfrak{Q}_{KG}$ is minimal projective if, and only if, it is a projective cover if, and only if, it is maximal essential; and any two such are isomorphic.

The last assertion is an immediate consequence of Proposition 7, §9.8, p.210.

The hypothesis on KG is equivalent to assuming KG is *semi-perfect* in the terminology of Bass [1]. (Recall that all modules in this chapter are finitely generated!)

In view of the Projective Cover Theorem (§9.8, p.207), KG is semi-perfect if K is I-complete and K/I has descending chain condition. For example, K could be a field, or $\mathbb{Z}_p$, the p-adic integers. (The case $K = \mathbb{F}_p$ of Theorem 1 was discovered by Gaschütz [2].)

Another case when Theorem 1 applies is when KG is a local ring: i.e., when G is a p-group and K is a local ring in which p is not invertible (Proposition 10, §10.5, p.239). For then, if C is any KG-module and we choose a free KG-module E with a basis (e_i) in one-one correspondence with a basis $(c_i + CJ)$ of of C/CJ ($J = \mathrm{Jac}(KG)$), then $\varphi: E \longrightarrow C$, defined by $e_i \longmapsto c_i$, is surjective (Corollary, p.98 applies as C is finitely generated) and is essential (as $\mathrm{Ker}\varphi \leq EJ \leq \mathrm{Fr}_{KG}(E)$: cf. example 4, p.97 and the example in §9.8, p.207). Thus φ is a projective cover of C.

As a matter of fact we shall see later (and directly) that more than Theorem 1 is true when KG is local: any minimal free extension is a projective cover (Theorem 8, §11.6).

To have the group algebra KG semi-perfect is by no means a necessary condition for the existence of projective covers in $\mathfrak{Q}_{KG}$.

<u>PROPOSITION 2. Let G be cyclic of order n.</u>
<u>(i) If $n \in \mathrm{Jac}(K)$, then a minimal free pair of $\mathfrak{Q}_{KG}$ is a projective cover.</u>
<u>(ii) If n is not a prime power and $K = \mathbb{Z}_{(G)}$, then (i) applies but KG is not semi-perfect.</u>

<u>Proof of (i).</u> Take $1 \longrightarrow R \longrightarrow F \longrightarrow G \longrightarrow 1$ with F infinite cyclic on x. Suppose there exists an epimorphism $\alpha : (\bar{R},\chi)_{(K)} \longrightarrow (C,0)$. If $\alpha = \Gamma(\alpha, \sigma)$, then $x\sigma = g^r c$, say (where $G = \langle g \rangle$, $c \in C$) and $x^n\alpha = x^n\sigma = nc$. But $(x^n\alpha)K = C$ and so $C = nC$. Thus $C = 0$ (Corollary on p.98) and hence $(\bar{R},\chi)_{(K)}$ is essential.

<u>Remark</u>. If $|G| = p$ and $K = \mathbb{Z}_{(p)} \cap \mathbb{Z}_{(q)}$, then $p \notin \mathrm{Jac}(K)$ and $F/R'R^q$ is a split image of a minimal free extension. Thus $nK \neq K$ is not a sufficient condition for the truth of (i).

Part (ii) of Proposition 2 is a consequence of

<u>PROPOSITION 3. Let K be a semi-local Dedekind domain of characteristic zero in which every prime dividing $|G|$ is non-invertible. If KG is semi-perfect, then G is a prime-power group.</u>

<u>Proof</u>. Let $p/|G|$ and $\mathfrak{p}$ be a prime of K containing p, so that $F = K/\mathfrak{p}$ is a field of characteristic p. Assume G is not a p-group. Then FG is not local (Proposition 10, §10.5, p.241) and so there exists an indecomposable projective but non-free FG-module M. Now M is also a KG-module via $K \to F$ and by hypothesis there is a projective cover $f : P \to M$ in Mod_{KG}. Then $P/P\mathfrak{p} \to M$ is a projective cover in Mod_{FG} and so $P/P\mathfrak{p} \cong M$ (uniqueness of projective covers). By Swan's Theorem (Theorem 2, §10.6, p.243), P is KG-free and hence $P/P\mathfrak{p}$ is FG-free. This contradicts the choice of M.

<u>Exercises</u>

1. Show that the converse of Proposition 3 is false.(Use the remark above.)

2. For any G, $\mathbb{Z}G$ is not semi-perfect.

<u>Problem</u>. Find necessary and sufficient conditions on K or KG

to ensure that $\mathfrak{Q}_{KG}$ contains a projective cover. (The related problem of necessary and sufficient conditions for KG to be semi-perfect has been studied by Sheila M. Kaye in her thesis (McGill University, 1969).)

11.3 Cohomological properties of projectives.

PROPOSITION 4. (Tate) Let (A,x) be a projective pair in $\mathfrak{Q}_{KG}$ and write $|G| = n$. Then

(i) $H^{q+2}(G,A) \underset{dd}{\overset{\sim}{\leftarrow}} H^q(G,K)$, all $q > 0$;

(ii) $H^2(G,A) \xrightarrow{\sim} K/nK$ (as K-modules) via $x \mapsto 1 + nK$;

(iii) $H^1(G,A) \underset{d}{\overset{\sim}{\leftarrow}} \mathfrak{g}^G$ (with $\mathfrak{g}$ the augmentation ideal of KG).

In (i) and (iii), d denotes an appropriate connecting homomorphism.

Proof. Let $(\bar{R},\chi)_{(K)}$ be a free pair so that

$$(\bar{R},\chi)_{(K)} = (A,x) \sqcap (P,0).$$

Since P is KG-projective and therefore has trivial cohomology, it will be sufficient to prove the proposition for the free pair $(\bar{R},\chi)_{(K)}$.

In the notation of §3.1 (p.31), but with K as coefficients, we have the short exact sequences

$$0 \to \mathfrak{g} \to KG \to K \to 0, \tag{1}$$

$$0 \to \mathfrak{r}/\mathfrak{f}\mathfrak{r} \to \mathfrak{f}/\mathfrak{f}\mathfrak{r} \to \mathfrak{g} \to 0. \tag{2}$$

The corresponding cohomology sequences give

$$H^q(G,K) \underset{d}{\xrightarrow{\sim}} H^{q+1}(G,\mathfrak{g}) \underset{d}{\xrightarrow{\sim}} H^{q+2}(G, \mathfrak{r}/\mathfrak{f}\mathfrak{r})$$

for all $q > 0$ and this is (i). (Recall that $\mathfrak{r}/\mathfrak{f}\mathfrak{r} \simeq \bar{R}_{(K)}$.)

Sequence (2) also yields

$$H^1(G, \mathfrak{g}) \underset{d}{\xrightarrow{\sim}} H^2(G, \mathfrak{K}/\mathfrak{f}\mathfrak{K})$$

via $\chi_1 \mapsto \chi_{(K)}$, where χ_1 is the cohomology class induced by the identity on $\mathfrak{g}$ (cf. §8.2, p.122); and sequence (1) yields

$$0 \to \mathfrak{g}^G \to (KG)^G \to K \to H^1(G, \mathfrak{g}) \to 0.$$

But $(KG)^G = K\tau$, where $\tau = \sum_{x \in G} x$ and so $(KG)^G$ has image nK in K. Now the connecting homomorphism d induces

$$K/nK \xrightarrow{\sim} H^1(G, \mathfrak{g})$$

and $1 + nK \mapsto \chi_1$. Hence we have (ii).

Finally, from (2),

$$0 \to (\mathfrak{K}/\mathfrak{f}\mathfrak{K})^G \to (\mathfrak{f}/\mathfrak{f}\mathfrak{K})^G \to \mathfrak{g}^G \to H^1(G, \mathfrak{K}/\mathfrak{f}\mathfrak{K}) \to 0$$

and this yields (iii) because $(\mathfrak{f}/\mathfrak{f}\mathfrak{K})^G \to \mathfrak{g}^G$ is the zero map:

$$(\mathfrak{f}/\mathfrak{f}\mathfrak{K})^G = \coprod_i K(1-x_i + \mathfrak{f}\mathfrak{K})\tau$$

and

$$(1-x_i^{\pi})\tau = 0.$$

<u>Remark</u>. If we had used the Tate cohomology of G (cf. §10.7, p. 245), then (i) of Proposition 4 holds for all q in $\mathbb{Z}$ and (ii), (iii) are special cases of (i).

<u>COROLLARY. If the additive group of K is torsion-free (or even merely n-torsion-free: i.e., $nk = 0$ implies $k = 0$), then</u>

$$\underline{H^1(G,A) = 0 = H^3(G,A).}$$

<u>Proof</u>. $(KG)^G \cap \mathfrak{g}$ is isomorphic to all k such that $nk = 0$; and

$$H^1(G,K) \cong \operatorname{Hom}(G,K) = 0$$

since $g^n = 1$ for all g in G.

Exercises.

1. If G_p is a Sylow p-subgroup of G, prove that $H^2(G, \bar{R}_{(\mathbb{Z}_p)})$ is cyclic of order $|G_p|$ and the restriction

$$H^2(G,\bar{R}_{(\mathbb{Z}_p)}) \to H^2(G_p,\bar{R}_{(\mathbb{Z}_p)})$$

is an isomorphism.

2. Prove that for any A in $\mathrm{Mod}_{\mathbb{Z}_pG}$, restriction is (i) injective but (ii) need not be surjective. (For (i) use Corollary 2 on p.91; for (ii) look at $G = \langle x,y;\ x^3=y^2,\ xyx=y\rangle$ over $\mathbb{Z}_3$ with A trivial (Cartan-Eilenberg, p.253).

One point in Proposition 4 still needs clearing up. It concerns part (iii). Since $K\tau \cap \mathfrak{g} = \mathfrak{g}^G$, we have $K\tau = \mathfrak{g}^G$ if, and only if, $\tau \in \mathfrak{g}$ if, and only if, $n = 0$ in K. This happens, for example, if K is a field of characteristic p dividing n. What is the image of τ (if $\tau \in \mathfrak{g}$) in $H^1(G,\bar{R}_{(K)})$? Diagram chasing shows that it is precisely the cohomology class containing the derivation

$$w^\pi \longmapsto \prod_i (t_{i(w)}{}^{-1} r_{i,w}\ t_{i(w)})R' \otimes 1,$$

where π is $F \to G$, (t_i) is a transversal of R in F and $t_i w = r_{i,w} t_{i(w)}$.

This prompts the following observations.

Given $(A|E)$ in $\left(\frac{G}{}\right)$, take a transversal $T = (t_i)$ of A in E and suppose $t_i\ e = a_{i,e} t_{i(e)}$ $(e \in E)$. Then

$$d_T : e \mapsto \prod_i (t_{i(e)}^{-1}\, a_{i,e}\, t_{i(e)})$$

is a derivation of E in A. (Note that $e \mapsto \prod_i a_{i,e}$ is the ordinary transfer homomorphism and that d_T actually coincides with this on A: $ad_T = a^n$ where $n = |G|$.)

If $S = (s_i)$ is a second transversal and $s_i = c_i t_i$ $(c_i \in A)$, then

$$ed_S = [c,e]\,(e\, d_T),$$

where $c = \prod_i t_i^{-1} c_i t_i$. Thus d_S is cohomologous to d_T. We therefore obtain a unique element ν of $H^1(E,A)$ that we propose to call the <u>transit class</u> of $(A|E)$ or of $\Gamma(A|E) = (A,x)$.

If $A^n = 1$ (which happens when $nK = 0$), then $A \leq \mathrm{Ker}\, d_T$ and so d_T is effectively a derivation of G in A. We then let ν also denote the resulting element in $H^1(G,A)$ and call <u>this</u> the transit class of the extension. It is easy to see that an epimorphism in $\binom{G}{-}$ or $\mathfrak{Q}_G$ carries transit class to transit class.

<u>It follows that if $\tau \in \mathfrak{M}$, the isomorphism of Proposition 4, (iii), maps τ to the transit class of (A,x).</u>

11.4 Cohomological characterisation of projectives.

Recall that if H is a subgroup of G and i: $H \hookrightarrow G$ is the inclusion, i" is a functor: $\mathfrak{Q}_{KG} \to \mathfrak{Q}_{KH}$. We shall call this functor <u>restriction</u> and write i" = res (§9.1, p.189).

<u>THEOREM 2. Let K be field of characteristic p dividing $|G|$, and G_p a Sylow p-subgroup of G. The following are equivalent:</u>

<u>(i) (A,x) is projective in $\mathfrak{Q}_{KG}$;</u>

<u>(ii) (A,x)res is projective in $\mathfrak{Q}_{KG_p}$;</u>

<u>(iii) $H^1(G_p,A) \cong K$ and the transit class $\neq 0$,</u>

<u>$H^2(G_p,A) \cong K$ and x res $\neq 0$.</u>

<u>Proof.</u> (i) ⇒ (ii) follows from Theorem 4(K), §9.7, p.205, and (ii) ⇒ (iii) from Proposition 4.

<u>(iii) ⇒ (i).</u> Choose an epimorphism: $(\bar{R}, \chi)_{(K)} \to (A,x)$ with $(\bar{R}, \chi)_{(K)}$ free in $\mathfrak{Q}_{KG}$. Then the exact sequence

$$0 \to P \to \bar{R}_{(K)} \to A \to 0$$

yields (writing $H^q(M)$ for $H^q(G_p,M)$)

$$H^1(\bar{R}_{(K)}) \to H^1(A) \to H^2(P) \to H^2(\bar{R}_{(K)}) \to H^2(A).$$

Now $H^1(\bar{R}_{(K)}) \to H^1(A)$ is an isomorphism because it maps the transit class of $(\bar{R}, \chi)_{(K)}$res to the transit class of (A,x)res and both groups are one dimensional vector spaces over K. For the same reason, $H^2(\bar{R}_{(K)}) \to H^2(A)$ is an isomorphism since $\chi_{(K)}$res $\mapsto x$ res.

Hence $H^2(P) = 0$ and so P is KG-projective (Theorem 4(i), §10.7, p.243). Therefore $\bar{R}_{(K)} \to A$ splits (Proposition 3(i), §10.1, p.224) and consequently (A,x) is projective in $\mathfrak{Q}_{KG}$ (Theorem 4(K), §9.7, p.205).

<u>THEOREM 3. (A,x) is projective in $\mathfrak{Q}_G$ if, and only if,</u>

<u>(i) A is $\mathbb{Z}$-free;</u>

<u>for all primes p and any Sylow p-subgroup G_p,</u>

<u>(ii) $H^1(G_p,A) = 0$, and</u>

<u>(iii) $H^2(G_p,A)$ is generated by x res and of order $|G_p|$.</u>

<u>Proof.</u> Proposition 4 and its Corollary yield half the result. Conversely, choose an epimorphism $\varphi: (\bar{R},\chi) \to (A,x)$. Then

$$\varphi^*: \quad H^2(G_p,\bar{R}) \to H^2(G_p,A)$$

is an isomorphism by (iii). If $P = \operatorname{Ker}\varphi$, we have the exact sequence (writing $H^q(M)$ for $H^q(G_p,M)$)

$$H^1(A) \to H^2(P) \to H^2(\bar{R}) \to H^2(A) \to H^3(P) \to H^3(\bar{R}).$$

We conclude $H^2(P) = 0$ (using (ii)) and $H^3(P) = 0$ (since $H^3(\bar{R}) = 0$ by the corollary to Proposition 4). But P is $\mathbb{Z}$-free and so P is $\mathbb{Z}G$-projective (Theorem 3, §10.7, p.242) and therefore φ is a split epimorphism (Proposition 3(i), §10.1, p.224) and (A,x) is projective.

Theorem 3 implies that <u>(A,x) is projective in $\mathfrak{Q}_G$ if, and only if, (A,x)res is projective in $\mathfrak{Q}_{G_p}$ for all p.</u>

A more interesting "localization result" is the following.

<u>THEOREM 4. (A,x) is projective in $\mathfrak{Q}_G$ if, and only if,</u>

<u>$(A,x)_{(\mathbb{Z}_{(p)})}$ is projective in $\mathfrak{Q}_{\mathbb{Z}_{(p)}G}$, for all p.</u>

<u>Proof.</u> If (A,x) is projective, then clearly so is $(A,x)_{(K)}$ for

any K.

Converse: Let us write A_p for $A_{(\mathbb{Z}_{(p)})} = A \otimes \mathbb{Z}_{(p)}$. Then A_p is torsion-free for all p and hence A is torsion-free, whence $\mathbb{Z}$-free (by finite generation).

If H is a subgroup of G, then the restriction functor $\mathcal{Q}_{KG} \to \mathcal{Q}_{KH}$ maps projectives to projectives. Hence, in particular, if G_q is a Sylow q-subgroup of G, $(A,x)_p$ restricts to a projective object in $\mathcal{Q}_{\mathbb{Z}_{(p)}G_q}$. Thus by Proposition 4, for all p, $H^1(G_q,A_p) = 0$ and $H^2(G_q,A_p)$ is cyclic on $x_p$res of order $|G_q|\delta_{pq}$. Now

$$H^k(G_q,A) = \coprod_p H^k(G_q,A_p),$$

(Proposition 9, §10.4, p.236). Hence $H^1(G_q,A) = 0$ and $H^2(G_q,A)$ is cyclic on x res of order $|G_q|$. So (A,x) is projective by Theorem 3.

The "only if" half of Theorem 4 is true even if non-finitely generated modules are allowed. It is worth observing however that the "if" part is then definitely false: Let

$$A = \{a/b;\ a,b \in \mathbb{Z} \text{ and } b \text{ square-free}\}$$

and set $B = \frac{1}{p}\mathbb{Z}$. Then A/B is a torsion p'-group and so $(A/B)_p = (A/B) \otimes \mathbb{Z}_{(p)} = 0$. Hence $0 \to B \to A \to A/B \to 0$ yields $B_p \cong A_p$. But $B \cong \mathbb{Z}$ so that $B_p \cong \mathbb{Z}_{(p)}$. Thus $A_p \cong \mathbb{Z}_{(p)}$ and this is true for every p. Yet A is clearly not $\mathbb{Z}$-free. Let G = 1, the trivial group. Then $(A,0)_{(\mathbb{Z}_{(p)})} = (\mathbb{Z}_{(p)},0)$ is projective in $\mathfrak{L}_{\mathbb{Z}_{(p)}1}$ for all p and yet (A,0) is not projective in $\mathfrak{L}_1$.

Exercises.

1. If A is a $\mathbb{Z}_{(p)}G$-module, prove that

$$H^k(G,A) \cong H^k(G,A \otimes \mathbb{Z}_p)$$

for all $k > 0$. ($\mathbb{Z}_p$ is $\mathbb{Z}_{(p)}$-flat: cf., e.g., Atiyah-Macdonald, Introduction to Commutative Algebra, p.109; then use Proposition 7 of §10.3, p.233.)

2. Prove that (A,x) is projective in $\mathfrak{L}_{\mathbb{Z}_{(p)}G}$ if, and only if, (i) A is $\mathbb{Z}_{(p)}$-free; (ii) $H^1(G_p,A) = 0$; (iii) $H^2(G_p,A)$ is generated by x res and of order $|G_p|$. (Imitate the proof of Theorem 3.)

3. Prove that (A,x) is projective in $\mathfrak{L}_{\mathbb{Z}_{(p)}G}$ if, and only if, $(A,x)_{(\mathbb{Z}_p)}$ is projective in $\mathfrak{L}_{\mathbb{Z}_pG}$. (Use exercises 1 and 2.)

11.5 Uniqueness of minimal projectives.

We have seen that in many good situations a projective cover exists and then automatically minimal projectives are unique to within isomorphism. However, the existence of a projective cover is not necessary for the uniqueness of minimal projectives. To establish this (cf. Proposition 6 below) we first prove a general result.

<u>THEOREM 5. Any two minimal projectives in $\mathfrak{D}_{\mathbb{Z}_{(G)}G}$ are isomorphic.</u>

<u>PROPOSITION 5. If (A_i, x_i), $i = 1,2$, are projective in $\mathfrak{D}_{KG}$, then there exist KG-projective modules P_1, P_2 so that $A_1 \oplus P_1 \cong A_2 \oplus P_2$.</u>

(This result does not depend on G being finite, or the modules being finitely generated and requires no hypothesis on K.)

<u>Proof</u>. In view of Theorem 4(K) of §9.7 (p.205) we only need prove the proposition for free pairs $(\overline{R}_i, \chi_i)_{(K)}$. For these, the result is immediate from the module version of Schanuel (Lemma 11, §8.10, p.162) applied to

$$0 \to \mathfrak{r}_i/\mathfrak{f}_i\mathfrak{r}_i \to \mathfrak{f}_i/\mathfrak{f}_i\mathfrak{r}_i \to \mathfrak{g} \to 0, \quad i = 1,2.$$

<u>Proof of Theorem 5</u>. Let (A_i, x_i) be two minimal projectives. By Proposition 5,

$$A_1 \oplus P_1 \cong A_2 \oplus P_2$$

for suitable $\mathbb{Z}_{(G)}G$-projective modules P_1, P_2. Now $\mathbb{Z}_{(G)}$ is a semi-local Dedekind domain in which all primes dividing $|G|$ are non-invertible. Moreover, A_1, A_2 have no projective direct summand (Corollary to Proposition 5, §9.6, p.203 and cf. p.206). Hence by the Corollary in §10.6 (p.242), $A_1 \xrightarrow{\sim} A_2$ by an isomorphism φ, say. Now φ may not be a morphism in $\mathfrak{D}_{\mathbb{Z}_{(G)}G}$ but at any rate φ induces an isomorphism $H^2(G, A_1) \xrightarrow{\sim} H^2(G, A_2)$. By Proposition 4 above (§11.3), $x_1\varphi^* = vx_2$, where v is invertible in $\mathbb{Z}/|G|\mathbb{Z}$. If u is an invertible element in $\mathbb{Z}_{(G)}$ with image v under $\mathbb{Z}_{(G)} \to \mathbb{Z}/|G|\mathbb{Z}$, then $\theta = \frac{1}{u}\varphi$ is still a module isomorphism: $A_1 \xrightarrow{\sim} A_2$ and $x_1\theta^* = x_2$. So θ: $(A_1,x_1) \xrightarrow{\sim} (A_2,x_2)$.

Whether Theorem 5 remains true with $\mathbb{Z}_{(G)}$ replaced by $\mathbb{Z}$ remains an open question. If this were so, then even for p-groups it would be extremely interesting (cf. in this connexion §7.2, especially p.103). At least we may assert the following:

<u>If (A_1,x_1), (A_2,x_2) are minimal projectives in $\mathfrak{L}_G$, then</u>

$$(A_1,x_1)_{(\mathbb{Z}_{(G)})} \cong (A_2,x_2)_{(\mathbb{Z}_{(G)})}.$$

This is an immediate consequence of Theorem 5 and

<u>THEOREM 6. If (A,x) is minimal projective in $\mathfrak{L}_G$, then $(A,x)_{(\mathbb{Z}_{(G)})}$ is minimal projective in $\mathfrak{L}_{\mathbb{Z}_{(G)}G}$.</u>

<u>Proof</u>. We shall write $A_{(G)}$ for $A_{(\mathbb{Z}_{(G)})} = A \otimes \mathbb{Z}_{(G)}$ and similarly $(A,x)_{(G)}$. The projectivity of $(A,x)_{(G)}$ is clear. Hence we need only prove that $A_{(G)}$ has no $\mathbb{Z}_{(G)}G$-projective direct summand. Moreover, $\mathbb{Z}_{(G)}G$-projective is the same as $\mathbb{Z}_{(G)}G$-free, by Swan's theorem (p.241).

Suppose $A_{(G)} = B \oplus C$, where C is $\mathbb{Z}_{(G)}G$-free on (c_i) and let D be $\mathbb{Z}G$-free on (c_i). Then $D_{(G)} \cong C$. Now $c_i = a_i \otimes \frac{1}{m_i}$, for some a_i in A and $(m_i,|G|) = 1$. If ψ: $D \to A$ is the module homomorphism defined by $c_i \mapsto a_i$, then ψ is one-one and $\psi_{(G)}$ maps $D_{(G)}$ onto C. Hence $(A/D\psi)_{(G)} \cong B$.

Let $T/D\psi$ be the torsion group of $A/D\psi$ and $E = A/T$. Then

$D_{(G)} \xrightarrow{\sim} T_{(G)}$ via $\phi_{(G)}$ because coker $\phi_{(G)} \cong B$ (as we have seen) and B is torsion-free. Thus $E_{(G)} \cong B$.

Now T is $\mathbb{Z}G$-projective by Corollary 3 to Proposition 9, §10.4, p.238. Since E is $\mathbb{Z}$-free, A splits over T by Proposition 3(i), §10.1, p.224. Hence T = 0 by the minimality of (A,x). Thus C = 0, as required.

PROPOSITION 6. If $G = C_2 \times C_6$ (C_r = cyclic group of order r), then $\mathfrak{Q}_{\mathbb{Z}_{(G)}G}$ contains no projective cover (but any two minimal projectives are isomorphic).

Proof. It follows from Theorem 7 in the next section that if $(\bar{R}|\bar{F})$ is a free extension in $\mathfrak{Q}_G$ with $d(F) = 2$ (the minimum rank possible) then $(\bar{R}|\bar{F})_{(G)}$ is minimal projective. (Notation here as explained at the beginning of the proof of Theorem 6.)

It will be sufficient therefore to show that $(\bar{R}|\bar{F})_{(G)}$ is not essential. We do this by exhibiting a split image, as follows.

Let $G = \langle g_1, g_2; \ g_1{}^2 = g_2{}^6 = 1\rangle$, V be two-dimensional over $\mathbb{F}_3$ with basis v_1, v_2 and make V an $\mathbb{F}_3G$-module by

$$g_1 \mapsto \begin{pmatrix} 0 & 1 \\ 1 & 0 \end{pmatrix}, \quad g_2 \mapsto \begin{pmatrix} 2 & 0 \\ 0 & 2 \end{pmatrix}$$

relative to v_1, v_2. In $E = V]G$, we calculate $[g_1v_1, g_2v_2] = v_2$, $(g_1v_1)^{-1}v_2(g_1v_1) = v_1$. Hence $\langle g_1v_1, g_2v_2\rangle$ contains V and so equals E.

If F is free on x_1, x_2, define φ: $F \to E$ by $x_i \mapsto g_iv_i$.

Then

$$\begin{array}{ccc} (\bar{R}|\bar{F}) & \xrightarrow{\bar{\varphi}\ (\mathrm{epi})} & (V|E) \\ \Big\downarrow{\scriptstyle i} & & \\ (\bar{R}|\bar{F})_{(G)} & & \end{array}$$

and this can be completed at the module level by

$$\psi:\ \ \bar{r} \otimes \frac{1}{m} \longrightarrow \bar{r}\,\bar{\varphi}\,(\tfrac{1}{m})',$$

where $(\ )'$ is $\mathbb{Z}_{(G)} \longrightarrow \mathbb{F}_3$. Clearly, ψ is a $\mathbb{Z}_{(G)}G$-homomorphism and surjective. So there exists an epimorphism $(\bar{R}|\bar{F})_{(G)} \longrightarrow (V|E)$ (Lemma 1, §9.3, p.193), as needed.

COROLLARY. Let G_3 be the Sylow 3-subgroup of $G = C_2 \times C_6$. The image of a minimal projective in $\mathfrak{Q}_{\mathbb{Z}_{(G)}G}$ under restriction is not minimal projective in $\mathfrak{Q}_{\mathbb{Z}_{(G)}G_3}$.

Proof. In the notation of the above proof, if T is the inverse image of G_3 in F,

$$(\bar{R}|\bar{F})_{(G)}\mathrm{res} = (\bar{R}|\bar{T})_{(G)}.$$

This cannot be minimal projective in $\mathfrak{Q}_{\mathbb{Z}_{(G)}G_3}$ for $\bar{R}$ has

$$B = \left\langle x_2^{-2i} x_1^{\,2} x_2^{\,2i};\ \ i = 0,1,2 \right\rangle$$

as a direct summand qua G_3-module and $B \cong \mathbb{Z}G_3$.

Exercise. If $F/R \cong G$ is cyclic and F is free, prove that $\bar{R} \cong \mathbb{Z} \oplus (\mathbb{Z}G)^k$, where $k+1 = \mathrm{rank}\ F$. Hence a free extension is minimal projective if, and only if, the extension is minimal free (= cyclic).

11.6 Minimal free extensions.

Recall that $d(G)$ means the minimum number of generators of a group G (p.98). We shall say that a free presentation $1 \to R \to F \to G \to 1$ in which $d(F) = d(G)$ is a minimal presentation. (If G is a p-group, this coincides with the definition on p.100.) A free object in $\left(\frac{KG}{}\right)$ or $\mathfrak{Q}_{KG}$ arising from a minimal presentation we call a minimal free object.

THEOREM 7. If $d(G) \leq 2$, then a minimal free extension in $\left(\frac{\mathbb{Z}_{(G)}G}{}\right)$ is minimal projective.

Note that (by Theorem 5) there is only one minimal projective in the situation of Theorem 7.

Proof. If $d(G) = 1$ (i.e., G cyclic), the result is implied by Proposition 2(i), §11.2. As a matter of fact, it is clearly true for any integral domain K as coefficient ring: for then $(\bar{R}|\bar{F})_{(K)}$ minimal free implies R cyclic, hence $R_{(K)} \cong K$ and so is indecomposable.

Assume henceforth $d(G) = 2$ and take a minimal presentation $1 \to R \to F \to G \to 1$. Then

$$d(R) = |G|(2 - 1) + 1 = |G| + 1. \qquad (1)$$

If $\bar{R}_{(G)}$ (notation as in the first line of the proof of Theorem 6) had a projective summand, this must be $\mathbb{Z}_{(G)}G$-free (Theorem 2, §10.6, p.241) and hence $\mathbb{Z}_{(G)}$-free of rank a multiple of $|G|$. By (1) we must have

$$\bar{R}_{(G)} \cong T \oplus \mathbb{Z}_{(G)}G, \qquad (2)$$

where T is $\mathbb{Z}_{(G)}$-free of rank one. By Proposition 8, §10.4 (p.234), $T \cong S_{(G)}$, for some $\mathbb{Z}G$-module S and where we may assume $S \cong \mathbb{Z}$ as additive groups. By Corollary 2(i), §10.4 (p.238), $H^k(G,T) \cong H^k(G,S)$. The projectivity of (T,x) implies (Proposition 4) that $H^1(G,S) = 0$ and $H^2(G,S)$ is cyclic of order $|G|$.

We distinguish two cases.

<u>(1)</u> The action of G on S is <u>not</u> trivial. Then there exists u in G such that $su = -s$ (for $|\text{Aut } S| = 2$). If H is the kernel of the representation of G on S and $d \in \text{Der}(G,S)$, d_H is a homomorphism and so $d_H = 0$ (as H is finite and S is torsion-free). Hence d is determined by ud and so $\text{Der}(G,S) \cong \mathbb{Z}$. Under this isomorphism, $\text{Ider}(G,S) \cong 2\mathbb{Z}$ and therefore $|H^1(G,S)| = 2$. Consequently this case cannot arise.

<u>(2)</u> S is trivial as G-module, i.e., $S \cong \mathbb{Z}$. Now $H^2(G,\mathbb{Z})$ is the cokernel of

$$\rho: \text{Hom}(F/F',\mathbb{Z}) \longrightarrow \text{Hom}(R/[R,F],\mathbb{Z})$$

(cf. Proposition 6, p.47) and $R/[R,F] \cong RF'/F' \oplus H$ where $H = H_2(G,\mathbb{Z})$ is finite. Hence

$$\text{Hom}(R/[R,F],\mathbb{Z}) \cong \text{Hom}(RF'/F',\mathbb{Z})$$

and so $\text{Coker } \rho \cong G/G'$. Thus $G' = 1$ and G is cyclic. But this is ruled out as $d(G) = 2$.

We conclude that $\bar{R}_{(G)}$ has no decomposition as in (2) above and hence the theorem is proved.

We leave as an open problem whether the restriction $d(G) \leq 2$ in Theorem 7 is really necessary. It is certainly not true that the 2-generator groups are theonly ones for which the conclusion of Theorem 7 is valid:

<u>THEOREM 8. If G is a p-group and K is a local ring in which p is not invertible, then a minimal free extension in $\left(\frac{KG}{}\right)$ is a projective cover (and so, in particular, is minimal projective).</u>

<u>Remark</u>. Our proof happens to yield the projective cover property directly. If we could only establish the minimal projectivity, then the stronger assertion would nevertheless follow from what we already know: cf. the remarks after Theorem 1 in §11.2.

<u>Proof</u>. Suppose $1 \to R \to F \to G \to 1$ is a minimal presentation. Then the corresponding minimal free extension $(\bar{R}|\bar{F})_{(\mathbb{F}_p)}$ in $\left(\frac{\mathbb{F}_p G}{p}\right)$ is a projective cover because $R/R'R^p$ lies in the Frattini group of the finite p-group $F/R'R^p$.

Suppose φ: $(\bar{R},\chi)_{(K)} \to (P,0)$ is an epimorphism in $\mathfrak{Q}_{KG}$. Without loss of generality we can assume $P\mathfrak{m} = 0$ (where $\mathfrak{m}$ is the maximal ideal of K). Thus $(p\bar{R})_{(K)}\varphi = 0$ so that φ yields ψ: $R/R'R^p \to P$ and we have the commutative square

$$\begin{array}{ccc} (\bar{R},\chi) & \longrightarrow & (\bar{R},\chi)_{(K)} \\ \downarrow & & \downarrow \varphi \\ (\bar{R},\chi)_{(\mathbb{F}_p)} & \xrightarrow{\psi} & (P,0) \end{array}$$

By the projective cover property of $(\bar{R},\chi)_{(\mathbb{F}_p)}$, $P = 0$. Thus $(\bar{R}|\bar{F})_{(K)}$ is a projective cover.

<u>Exercise</u>. If G is such that minimal free = minimal projective when the coefficient ring is $\mathbb{Z}_{(G)}$, prove that every minimal free is minimal projective when the coefficient ring is $\mathbb{Z}$.

11.7 The module structure of minimal projectives.

Let (A,x) be a minimal projective in $\mathfrak{Q}_{KG}$. What can be said about the module-theoretic structure of A? This is an interesting but rather hard question. One general fact can be proved.

<u>THEOREM 9. Assume KG has the property of Proposition 4 of §10.2 (p.227) and that K is a Heller module. If (A,x) is a minimal projective in $\mathfrak{Q}_{KG}$, then A is indecomposable as KG-module.</u>

In the presence of Proposition 4, §10.2, K is Heller precisely when K' is indecomposable and $|G|K \neq K$ (Proposition 21, p.176). Thus Theorem 9 applies in particular if K is an integral domain, $|G|K \neq K$ and the Krull-Schmidt theorem holds for KG. If also applies if $K = \mathbb{Z}_{(G)}$: cf. the Corollary in §10.6.

<u>Proof</u>. Since (A,x) is minimal projective, $A \neq 0$ and A = A' in the notation of Proposition 4, §10.2. Hence, by that Proposition, A is indecomposable if, and only if, A is a Heller module.

Suppose $(\bar{R},x)_{(K)} = (A,x) \oplus (P,0)$. Then A is Heller if, and only if, $\bar{R}_{(K)}$ is Heller and we have, as usual,

$$0 \to \bar{R}_{(K)} \to \mathfrak{f}/\mathfrak{f}\mathfrak{r} \to KG \to K \to 0.$$

Since K is Heller, so is $\bar{R}_{(K)}$ by Proposition 5, §10.2.

<u>COROLLARY. If (A,x) is minimal projective in $\mathfrak{Q}_G$, then A is an indecomposable G-module.</u>

<u>Proof</u>. By Theorem 6, $(A,x)_{(G)}$ is minimal projective in $\mathfrak{Q}_{\mathbb{Z}_{(G)}G}$ and by Theorem 9, $A_{(G)} = A \otimes \mathbb{Z}_{(G)}$ is indecomposable. Hence A is indecomposable.

<u>Exercises</u>.

1. If G is a p-group and $1 \to R \to F \to G \to 1$ is a minimal presentation then R/R' is indecomposable as G-module. (Use Theorem 8 and either Theorem 9 with $K = \mathbb{F}_p$ or the exercise immediately after Theorem 8 and the above Corollary.)

2. Let G be a p-group and $K = \mathbb{F}_p \oplus \mathbb{F}_p$. Then K is not a Heller module.

Let (A,x) be minimal projective in $\mathfrak{Q}_{\mathbb{F}_pG}$. Gaschütz showed in [2] how the structure of A can be described in terms of the group algebra $\mathbb{F}_pG$. His result can be expressed in the following more general form.

THEOREM 10. (Gaschütz). Assume KG is semi-perfect (i.e., all - finitely generated! - modules have projective covers) and that the condition of Proposition 4, §10.2 holds. Let

$$V \xrightarrow{\varphi} U \xrightarrow{\psi} K \longrightarrow 0$$

be exact with ψ a projective cover of K, φ a projective cover of $\operatorname{Ker}\psi$. If $(\bar{R}|\bar{F})$ is a free extension with $d = d(F)$, then

$$\bar{R}_{(K)} \oplus V \oplus KG \cong \operatorname{Ker}\varphi \oplus (KG)^d \oplus U.$$

When $|G|K = K$, this reduces to $\bar{R}_{(K)} \oplus KG \cong (KG)^d \oplus K$.

When $|G|K \neq K$, $\operatorname{Ker}\varphi$ is the module of the (unique) minimal projective in $\mathfrak{Q}_{KG}$.

Proof. Apply Lemma 12, p.163, to the sequences

$$0 \to \bar{R}_{(K)} \to \mathfrak{f}/\mathfrak{f}\mathfrak{r} \to KG \to K \to 0$$

$$0 \to \operatorname{Ker}\varphi \to V \to U \to K \to 0.$$

For the last part, we note that if $\varphi\colon V \to B$ is a projective cover with $B \in \operatorname{Lat}_{KG}$, then $W = \operatorname{Ker}\varphi$ has no non-zero projective summand. For suppose $W = C \oplus P$, with P KG-projective. Since B and V are K-projective, so is W and hence so is V/P. By Proposition 3(i), §10.1, $V = V_0 \oplus P$ and thus $V = V_0$ by the projective cover property. Hence $P = 0$.

Note that when the Krull-Schmidt theorem holds for KG, the formula of Theorem 10 gives a reasonably explicit decomposition of $\bar{R}_{(K)}$.

11.8 Conclusion.

The problems concerning minimal projectives fall naturally into three classes. First there are the comparison type problems: Are any two minimal projectives isomorphic? If not, how can one classify the isomorphism classes? What happens to minimal projectives under various restriction maps?

Secondly, questions concerning the group-theoretic structure of minimal projectives: Are the minimal projectives essential? Are they free? How can they be recognised purely group theoretically?

Thirdly, there are the module-theoretic problems: How can minimal projectives be characterised cohomologically? What modules can arise in minimal projectives? How do minimal projectives behave under coefficient changes?

Perhaps the most interesting coefficient ring is $\mathbb{Z}$. But this is also, in many ways, the most difficult. A sensible coefficient ring, but still close to $\mathbb{Z}$, is $\mathbb{Z}_{(G)}$ (for given G). Here, as we have seen, some of the problems enumerated above can be solved satisfactorily.

We also found it necessary at times to restrict the type of group G considered. For example, p-groups turned out to be much easier to deal with, though even there some fascinating problems remain: cf. p.103, for example. In this connexion we state one final result: if $(\overline{R}|\overline{F})$ is a minimal free extension in $\left(\underline{G}\right)$, then

$$d_G(\overline{R}) = \dim H^2(G, \mathbb{F}_p) = d_G(\overline{R} \otimes \mathbb{F}_p).$$

Sources and references.

Theorem 1, §11.2, and Theorem 10, §11.7, were discovered by Gaschütz [2] when K is $\mathbb{F}_p$ (or, more generally, any field of characteristic p). The paper of Gaschütz was an important factor in the development of the present theory.

Proposition 4, §11.3, is due to J. Tate when K = $\mathbb{Z}$ and may be found in Kawada's paper [5].

Alex Heller showed me the example at the end of §11.4 in July 1968.

Irving Reiner showed me the crux of the proof of Theorem 6 in June 1967. It may now be read off from Theorem 3.3 in Jacobinski's remarkable paper [3]. (It seems quite likely that further applications are possible of the results in [3] to the theory of the present chapter.)

In connexion with §11.7 we mention that Heller's work on indecomposable modules has also recently been used to good effect by D.L. Johnson: cf., e.g., [4].

The theorem stated at the end of §11.8 will appear (I hope) in a separate publication.

[1] Bass, H.: Finitistic dimension and a homological generalisation of semi-primary rings, Trans. American Math. Soc. 95 (1960) 466-488.

[2] Gaschütz, W.: Über modulare Darstellungen endlicher Gruppen die von freien Gruppen induziert werden, Math.Zeit. 60 (1954) 274-286.

[3] Jacobinski, H.: Genera and decompositions of lattices over orders, Acta Math. 121 (1968) 1-29.

[4] Johnson, D.L.: Indecomposable representations of the group (p,p) over fields of characteristic p, J. London Math. Soc. (2) 1 (1969) 43-50.

[5] Kawada, Y.: Cohomology of group extensions, J.Fac. Sci. Tokyo Univ. 9 (1963) 417-431.

Offsetdruck: Julius Beltz, Weinheim/Bergstr.

Lecture Notes in Mathematics

Bisher erschienen/Already published

Vol. 1: J. Wermer, Seminar über Funktionen-Algebren. IV, 30 Seiten. 1964. DM 3,80 / $ 1.10

Vol. 2: A. Borel, Cohomologie des espaces localement compacts d'après. J. Leray. IV, 93 pages. 1964. DM 9,– / $ 2.60

Vol. 3: J. F. Adams, Stable Homotopy Theory. Third edition. IV, 78 pages. 1969. DM 8,– / $ 2.20

Vol. 4: M. Arkowitz and C. R. Curjel, Groups of Homotopy Classes. 2nd. revised edition. IV, 36 pages. 1967. DM 4,80 / $ 1.40

Vol. 5: J.-P. Serre, Cohomologie Galoisienne. Troisième édition. VIII, 214 pages. 1965. DM 18,– / $ 5.00

Vol. 6: H. Hermes, Term Logic with Choise Operator. III, 55 pages. 1970. DM 6,– / $ 1.70

Vol. 7: Ph. Tondeur, Introduction to Lie Groups and Transformation Groups. Second edition. VIII, 176 pages. 1969. DM 14, / $ 3.80

Vol. 8: G. Fichera, Linear Elliptic Differential Systems and Eigenvalue Problems. IV, 176 pages. 1965. DM 13,50 / $ 3.80

Vol. 9: P. L. Ivănescu, Pseudo-Boolean Programming and Applications. IV, 50 pages. 1965. DM 4,80 / $ 1.40

Vol. 10: H. Lüneburg, Die Suzukigruppen und ihre Geometrien. VI, 111 Seiten. 1965. DM 8,– / $ 2.20

Vol. 11: J.-P. Serre, Algèbre Locale. Multiplicités. Rédigé par P. Gabriel. Seconde édition. VIII, 192 pages. 1965. DM 12,– / $ 3.30

Vol. 12: A. Dold, Halbexakte Homotopiefunktoren. II, 157 Seiten. 1966. DM 12,– / $ 3.30

Vol. 13: E. Thomas, Seminar on Fiber Spaces. IV, 45 pages. 1966. DM 4,80 / $ 1.40

Vol. 14: H. Werner, Vorlesung über Approximationstheorie. IV, 184 Seiten und 12 Seiten Anhang. 1966. DM 14,– / $ 3.90

Vol. 15: F. Oort, Commutative Group Schemes. VI, 133 pages. 1966. DM 9,80 / $ 2.70

Vol. 16: J. Pfanzagl and W. Pierlo, Compact Systems of Sets. IV, 48 pages. 1966. DM 5,80 / $ 1.60

Vol. 17: C. Müller, Spherical Harmonics. IV, 46 pages. 1966. DM 5,– / $ 1.40

Vol. 18: H.-B. Brinkmann und D. Puppe, Kategorien und Funktoren. XII, 107 Seiten. 1966. DM 8,– / $ 2.20

Vol. 19: G. Stolzenberg, Volumes, Limits and Extensions of Analytic Varieties. IV, 45 pages. 1966. DM 5,40 / $ 1.50

Vol. 20: R. Hartshorne, Residues and Duality. VIII, 423 pages. 1966. DM 20,– / $ 5.50

Vol. 21: Seminar on Complex Multiplication. By A. Borel, S. Chowla, C. S. Herz, K. Iwasawa, J.-P. Serre. IV, 102 pages. 1966. DM 8,– / $ 2.20

Vol. 22: H. Bauer, Harmonische Räume und ihre Potentialtheorie. IV, 175 Seiten. 1966. DM 14,– / $ 3.90

Vol. 23: P. L. Ivănescu and S. Rudeanu, Pseudo-Boolean Methods for Bivalent Programming. 120 pages. 1966. DM 10,– / $ 2.80

Vol. 24: J. Lambek, Completions of Categories. IV, 69 pages. 1966. DM 6,80 / $ 1.90

Vol. 25: R. Narasimhan, Introduction to the Theory of Analytic Spaces. IV, 143 pages. 1966. DM 10,– / $ 2.80

Vol. 26: P.-A. Meyer, Processus de Markov. IV, 190 pages. 1967. DM 15,– / $ 4.20

Vol. 27: H. P. Künzi und S. T. Tan, Lineare Optimierung großer Systeme. VI, 121 Seiten. 1966. DM 12,– / $ 3.30

Vol. 28: P. E. Conner and E. E. Floyd, The Relation of Cobordism to K-Theories. VIII, 112 pages. 1966. DM 9,80 / $ 2.70

Vol. 29: K. Chandrasekharan, Einführung in die Analytische Zahlentheorie. VI, 199 Seiten. 1966. DM 16,80 / $ 4.70

Vol. 30: A. Frölicher and W. Bucher, Calculus in Vector Spaces without Norm. X, 146 pages. 1966. DM 12,– / $ 3.30

Vol. 31: Symposium on Probability Methods in Analysis. Chairman. D. A. Kappos. IV, 329 pages. 1967. DM 20,– / $ 5.50

Vol. 32: M. André, Méthode Simpliciale en Algèbre Homologique et Algèbre Commutative. IV, 122 pages. 1967. DM 12,– / $ 3.30

Vol. 33: G. I. Targonski, Seminar on Functional Operators and Equations. IV, 110 pages. 1967. DM 10,– / $ 2.80

Vol. 34: G. E. Bredon, Equivariant Cohomology Theories. VI, 64 pages. 1967. DM 6,80 / $ 1.90

Vol. 35: N. P. Bhatia and G. P. Szegö, Dynamical Systems. Stability Theory and Applications. VI, 416 pages. 1967. DM 24,– / $ 6.60

Vol. 36: A. Borel, Topics in the Homology Theory of Fibre Bundles. VI, 95 pages. 1967. DM 9,– / $ 2.50

Vol. 37: R. B. Jensen, Modelle der Mengenlehre. X, 176 Seiten. 1967. DM 14,– / $ 3.90

Vol. 38: R. Berger, R. Kiehl, E. Kunz und H.-J. Nastold, Differentialrechnung in der analytischen Geometrie IV, 134 Seiten. 1967 DM 12,– / $ 3.30

Vol. 39: Séminaire de Probabilités I. II, 189 pages. 1967. DM 14,– / $ 3.90

Vol. 40: J. Tits, Tabellen zu den einfachen Lie Gruppen und ihren Darstellungen. VI, 53 Seiten. 1967. DM 6.80 / $ 1.90

Vol. 41: A. Grothendieck, Local Cohomology. VI, 106 pages. 1967. DM 10,– / $ 2.80

Vol. 42: J. F. Berglund and K. H. Hofmann, Compact Semitopological Semigroups and Weakly Almost Periodic Functions. VI, 160 pages. 1967. DM 12,– / $ 3.30

Vol. 43: D. G. Quillen, Homotopical Algebra. VI, 157 pages. 1967. DM 14,– / $ 3.90

Vol. 44: K. Urbanik, Lectures on Prediction Theory. IV, 50 pages. 1967. DM 5,80 / $ 1.60

Vol. 45: A. Wilansky, Topics in Functional Analysis. VI, 102 pages. 1967. DM 9,60 / $ 2.70

Vol. 46: P. E. Conner, Seminar on Periodic Maps. IV, 116 pages. 1967. DM 10,60 / $ 3.00

Vol. 47: Reports of the Midwest Category Seminar I. IV, 181 pages. 1967. DM 14,80 / $ 4.10

Vol. 48: G. de Rham, S. Maumary et M. A. Kervaire, Torsion et Type Simple d'Homotopie. IV, 101 pages. 1967. DM 9,60 / $ 2.70

Vol. 49: C. Faith, Lectures on Injective Modules and Quotient Rings. XVI, 140 pages. 1967. DM 12,80 / $ 3.60

Vol. 50: L. Zalcman, Analytic Capacity and Rational Approximation. VI, 155 pages. 1968. DM 13.20 / $ 3.70

Vol. 51: Séminaire de Probabilités II. IV, 199 pages. 1968. DM 14,– / $ 3.90

Vol. 52: D. J. Simms, Lie Groups and Quantum Mechanics. IV, 90 pages. 1968. DM 8,– / $ 2.20

Vol. 53: J. Cerf, Sur les difféomorphismes de la sphère de dimension trois (Γ_4 = O). XII, 133 pages. 1968. DM 12,– / $ 3.30

Vol. 54: G. Shimura, Automorphic Functions and Number Theory. VI, 69 pages. 1968. DM 8,– / $ 2.20

Vol. 55: D. Gromoll, W. Klingenberg und W. Meyer, Riemannsche Geometrie im Großen. VI, 287 Seiten. 1968. DM 20,– / $ 5.50

Vol. 56: K. Floret und J. Wloka, Einführung in die Theorie der lokalkonvexen Räume. VIII, 194 Seiten. 1968. DM 16,– / $ 4.40

Vol. 57: F. Hirzebruch und K. H. Mayer, O (n)-Mannigfaltigkeiten, exotische Sphären und Singularitäten. IV, 132 Seiten. 1968. DM 10,80 / $ 3.00

Vol. 58: Kuramochi Boundaries of Riemann Surfaces. IV, 102 pages. 1968. DM 9,60 / $ 2.70

Vol. 59: K. Jänich, Differenzierbare G-Mannigfaltigkeiten. VI, 89 Seiten. 1968. DM 8,– / $ 2.20

Vol. 60: Seminar on Differential Equations and Dynamical Systems. Edited by G. S. Jones. VI, 106 pages. 1968. DM 9,60 / $ 2.70

Vol. 61: Reports of the Midwest Category Seminar II. IV, 91 pages. 1968. DM 9,60 / $ 2.70

Vol. 62: Harish-Chandra, Automorphic Forms on Semisimple Lie Groups X, 138 pages. 1968. DM 14, / $ 3.90

Vol. 63: F. Albrecht, Topics in Control Theory. IV, 65 pages. 1968. DM 6,80 / $ 1.90

Vol. 64: H. Berens, Interpolationsmethoden zur Behandlung von Approximationsprozessen auf Banachräumen. VI, 90 Seiten. 1968. DM 8,– / $ 2.20

Vol. 65: D. Kölzow, Differentiation von Maßen. XII, 102 Seiten. 1968. DM 8,– / $ 2.20

Vol. 66: D. Ferus, Totale Absolutkrümmung in Differentialgeometrie und -topologie. VI, 85 Seiten. 1968. DM 8,– / $ 2.20

Vol. 67: F. Kamber and P. Tondeur, Flat Manifolds. IV, 53 pages. 1968. DM 5,80 / $ 1.60

Vol. 68: N. Boboc et P. Mustață, Espaces harmoniques associés aux opérateurs différentiels linéaires du second ordre de type elliptique. VI, 95 pages. 1968. DM 8,60 / $ 2.40

Vol. 69: Seminar über Potentialtheorie. Herausgegeben von H. Bauer. VI, 180 Seiten. 1968. DM 14,80 / $ 4.10

Vol. 70: Proceedings of the Summer School in Logic. Edited by M. H. Löb. IV, 331 pages. 1968. DM 20,– / $ 5.50

Vol. 71: Séminaire Pierre Lelong (Analyse), Année 1967 – 1968. VI, 19 pages. 1968. DM 14,– / $ 3.90

Bitte wenden / Continued

Vol. 72: The Syntax and Semantics of Infinitary Languages. Edited by J. Barwise. IV, 268 pages. 1968. DM 18,– / $ 5.00

Vol. 73: P. E. Conner, Lectures on the Action of a Finite Group. IV, 123 pages. 1968. DM 10,– / $ 2.80

Vol. 74: A. Fröhlich, Formal Groups. IV, 140 pages. 1968. DM 12,– / $ 3.30

Vol. 75: G. Lumer, Algèbres de fonctions et espaces de Hardy. VI, 80 pages. 1968. DM 8,– / $ 2.20

Vol. 76: R. G. Swan, Algebraic K-Theory. IV, 262 pages. 1968. DM 18,– / $ 5.00

Vol. 77: P.-A. Meyer, Processus de Markov: la frontière de Martin. IV, 123 pages. 1968. DM 10,– / $ 2.80

Vol. 78: H. Herrlich, Topologische Reflexionen und Coreflexionen. XVI, 166 Seiten. 1968. DM 12,– / $ 3.30

Vol. 79: A. Grothendieck, Catégories Cofibrées Additives et Complexe Cotangent Relatif. IV, 167 pages. 1968. DM 12,– / $ 3.30

Vol. 80: Seminar on Triples and Categorical Homology Theory. Edited by B. Eckmann. IV, 398 pages. 1969. DM 20,– / $ 5.50

Vol. 81: J.-P. Eckmann et M. Guenin, Méthodes Algébriques en Mécanique Statistique. VI, 131 pages. 1969. DM 12,– / $ 3.30

Vol. 82: J. Wloka, Grundräume und verallgemeinerte Funktionen. VIII, 131 Seiten. 1969. DM 12,– / $ 3.30

Vol. 83: O. Zariski, An Introduction to the Theory of Algebraic Surfaces. IV, 100 pages. 1969. DM 8,– / $ 2.20

Vol. 84: H. Lüneburg, Transitive Erweiterungen endlicher Permutationsgruppen. IV, 119 Seiten. 1969. DM 10,– / $ 2.80

Vol. 85: P. Cartier et D. Foata, Problèmes combinatoires de commutation et réarrangements. IV, 88 pages. 1969. DM 8,– / $ 2.20

Vol. 86: Category Theory, Homology Theory and their Applications I. Edited by P. Hilton. VI, 216 pages. 1969. DM 16,– / $ 4.40

Vol. 87: M. Tierney, Categorical Constructions in Stable Homotopy Theory. IV, 65 pages. 1969. DM 6,– / $ 1.70

Vol. 88: Séminaire de Probabilités III. IV, 229 pages. 1969. DM 18,– / $ 5.00

Vol. 89: Probability and Information Theory. Edited by M. Behara, K. Krickeberg and J. Wolfowitz. IV, 256 pages. 1969. DM 18,– / $ 5.00

Vol. 90: N. P. Bhatia and O. Hajek, Local Semi-Dynamical Systems. II, 157 pages. 1969. DM 14,– / $ 3.90

Vol. 91: N. N. Janenko, Die Zwischenschrittmethode zur Lösung mehrdimensionaler Probleme der mathematischen Physik. VIII, 194 Seiten. 1969. DM 16,80 / $ 4.70

Vol. 92: Category Theory, Homology Theory and their Applications II. Edited by P. Hilton. V, 308 pages. 1969. DM 20,– / $ 5.50

Vol. 93: K. R. Parthasarathy, Multipliers on Locally Compact Groups. III, 54 pages. 1969. DM 5,60 / $ 1.60

Vol. 94: M. Machover and J. Hirschfeld, Lectures on Non-Standard Analysis. VI, 79 pages. 1969. DM 6,– / $ 1.70

Vol. 95: A. S. Troelstra, Principles of Intuitionism. II, 111 pages. 1969. DM 10,– / $ 2.80

Vol. 96: H.-B. Brinkmann und D. Puppe, Abelsche und exakte Kategorien, Korrespondenzen. V, 141 Seiten. 1969. DM 10,– / $ 2.80

Vol. 97: S. O. Chase and M. E. Sweedler, Hopf Algebras and Galois theory. II, 133 pages. 1969. DM 10,– / $ 2.80

Vol. 98: M. Heins, Hardy Classes on Riemann Surfaces. III, 106 pages. 1969. DM 10,– / $ 2.80

Vol. 99: Category Theory, Homology Theory and their Applications III. Edited by P. Hilton. IV, 489 pages. 1969. DM 24,– / $ 6.60

Vol. 100: M. Artin and B. Mazur, Etale Homotopy. II, 196 Seiten. 1969. DM 12,– / $ 3.30

Vol. 101: G. P. Szegö et G. Treccani, Semigruppi di Trasformazioni Multivoche. VI, 177 pages. 1969. DM 14,– / $ 3.90

Vol. 102: F. Stummel, Rand- und Eigenwertaufgaben in Sobolewschen Räumen. VIII, 386 Seiten. 1969. DM 20,– / $ 5.50

Vol. 103: Lectures in Modern Analysis and Applications I. Edited by C. T. Taam. VII, 162 pages. 1969. DM 12,– / $ 3.30

Vol. 104: G. H. Pimbley, Jr., Eigenfunction Branches of Nonlinear Operators and their Bifurcations. II, 128 pages. 1969. DM 10,– / $ 2.80

Vol. 105: R. Larsen, The Multiplier Problem. VII, 284 pages. 1969. DM 18,– / $ 5.00

Vol. 106: Reports of the Midwest Category Seminar III. Edited by S. Mac Lane. III, 247 pages. 1969. DM 16,– / $ 4.40

Vol. 107: A. Peyerimhoff, Lectures on Summability. III, 111 pages. 1969. DM 8,– / $ 2.20

Vol. 108: Algebraic K-Theory and its Geometric Applications. Edited by R. M.F. Moss and C. B. Thomas. IV, 86 pages. 1969. DM 6,– / $ 1.70

Vol. 109: Conference on the Numerical Solution of Differential Equations. Edited by J. Ll. Morris. VI, 275 pages. 1969. DM 18,– / $ 5.00

Vol. 110: The Many Facets of Graph Theory. Edited by G. Chartrand and S. F. Kapoor. VIII, 290 pages. 1969. DM 18,– / $ 5.00

Vol. 111: K. H. Mayer, Relationen zwischen charakteristischen Zahlen. III, 99 Seiten. 1969. DM 8,– / $ 2.20

Vol. 112: Colloquium on Methods of Optimization. Edited by N. N. Moiseev. IV, 293 pages. 1970. DM 18,– / $ 5.00

Vol. 113: R. Wille, Kongruenzklassengeometrien. III, 99 Seiten. 1970. DM 8,– / $ 2.20

Vol. 114: H. Jacquet and R. P. Langlands, Automorphic Forms on GL (2). VII, 548 pages. 1970. DM 24,– / $ 6.60

Vol. 115: K. H. Roggenkamp and V. Huber-Dyson, Lattices over Orders I. XIX, 290 pages. 1970. DM 18,– / $ 5.00

Vol. 116: Séminaire Pierre Lelong (Analyse) Année 1969. IV, 195 pages. 1970. DM 14,– / $ 3.90

Vol. 117: Y. Meyer, Nombres de Pisot, Nombres de Salem et Analyse Harmonique. 63 pages. 1970. DM 6,– / $ 1.70

Vol. 118: Proceedings of the 15th Scandinavian Congress, Oslo 1968. Edited by K. E. Aubert and W. Ljunggren. IV, 162 pages. 1970. DM 12,– / $ 3.30

Vol. 119: M. Raynaud, Faisceaux amples sur les schémas en groupes et les espaces homogènes. III, 219 pages. 1970. DM 14,– / $ 3.90

Vol. 120: D. Siefkes, Büchi's Monadic Second Order Successor Arithmetic. XII, 130 Seiten. 1970. DM 12,– / $ 3.30

Vol. 121: H. S. Bear, Lectures on Gleason Parts. III, 47 pages. 1970. DM 6,–/$ 1.70

Vol. 122: H. Zieschang, E. Vogt und H.-D. Coldewey, Flächen und ebene diskontinuierliche Gruppen. VIII, 203 Seiten. 1970. DM 16,– / $ 4.40

Vol. 123: A. V. Jategaonkar, Left Principal Ideal Rings. VI, 145 pages. 1970. DM 12,– / $ 3.30

Vol. 124: Séminare de Probabilités IV. Edited by P. A. Meyer. IV, 282 pages. 1970. DM 20,– / $ 5.50

Vol. 125: Symposium on Automatic Demonstration. V, 310 pages. 1970. DM 20,– / $ 5.50

Vol. 126: P. Schapira, Théorie des Hyperfonctions. XI, 157 pages. 1970. DM 14,– / $ 3.90

Vol. 127: I. Stewart, Lie Algebras. IV, 97 pages. 1970 DM 10,– / $ 2.80

Vol. 128: M. Takesaki, Tomita's Theory of Modular Hilbert Algebras and its Applications. II, 123 pages. 1970. DM 10,– / $ 2.80

Vol. 129: K. H. Hofmann, The Duality of Compact Semigroups and C*-Bigebras. XII, 142 pages. 1970. DM 14,– / $ 3.90

Vol. 130: F. Lorenz, Quadratische Formen über Körpern. II, 77 Seiten. 1970. DM 8,– / $ 2.20

Vol. 131: A Borel et al., Seminar on Algebraic Groups and Related Finite Groups. VII, 321 pages. 1970. DM 22,– / $ 6.10

Vol. 132: Symposium on Optimization. III, 348 pages. 1970. DM 22,– / $ 6.10

Vol. 133: F. Topsøe, Topology and Measure. XIV, 79 pages. 1970. DM 8,– / $ 2.20

Vol. 134: L. Smith, Lectures on the Eilenberg-Moore Spectral Sequence. VII, 142 pages. 1970. DM 14,– / $ 3.90

Vol. 135: W. Stoll, Value Distribution of Holomorphic Maps into Compact Complex Manifolds. II, 267 pages. 1970. DM 18,– / $

Vol. 136: M. Karoubi et al., Séminaire Heidelberg-Saarbrücken-Strasbourg sur la K-Théorie. IV, 264 pages. 1970. DM 18,– / $ 5 00

Vol. 137: Reports of the Midwest Category Seminar IV. Edited by S. MacLane. III, 139 pages. 1970. DM 12,– / $ 3.30

Vol. 138: D. Foata et M. Schützenberger, Théorie Géométrique des Polynômes Eulériens. V, 94 pages. 1970. DM 10,– / $ 2 80

Vol. 139: A. Badrikian, Séminaire sur les Fonctions Aléatoires Linéaires et les Mesures Cylindriques. VII, 221 pages. 1970. DM 18,– / $ 5.00

Vol. 140: Lectures in Modern Analysis and Applications II. Edited by C. T. Taam. VI, 119 pages. 1970. DM 10,– / $ 2.80

Vol. 141: G. Jameson, Ordered Linear Spaces. XV, 194 pages. 1970. DM 16,– / $ 4.40

Vol. 142: K. W. Roggenkamp, Lattices over Orders II. V, 388 pages. 1970. DM 22,– / $ 6.10

Vol. 143: K. W. Gruenberg, Cohomological Topics in Group Theory. XIV, 275 pages. 1970. DM 20,– / $ 5.50